Greener Pastures
On Your Side Of The Fence

ABOUT THE AUTHOR

Bill grew up on a dairy farm in Wisconsin, obtained a B.S. in Zoology from the University of Wisconsin, and served as a Peace Corps Volunteer in Chile. After his Peace Corps service, he returned to graduate school at Wisconsin with his new Chilean wife, Lita, and obtained an M.S. in Soil Science and a Ph.D. in Agronomy. He did part of his M.S. research and all of his Ph.D. research in Rio Grande do Sul, Brasil. While there he heard about Voisin grazing management, but never saw it being used successfully. After completing the degree requirements, they returned to Rio Grande do Sul to work on a United Nations project. After that, Bill and Lita moved with their daughters, Michelle and Nicole, to Oregon where they lived while Bill worked on an experiment station for Oregon State University.

Bill and Lita now farm 25 acres of land in the Green Mountain foothills of Vermont's Champlain Valley. It was here in 1981, while grazing dairy heifers and trying the Voisin grazing management method, that Bill realized the dramatic improvements in plant and animal production that result from using it. By 1986 the productivity of their pasture had quadrupled, with no other inputs besides observation, Voisin management intensive grazing, and fencing. Bill also works at the University of Vermont teaching courses on pasture management, forage production, weed/crop ecology, and holistic management, and researching ways of improving pasture performance to cut production costs.

Greener Pastures On Your Side Of The Fence

Better Farming With
Voisin Management Intensive Grazing

by Bill Murphy

Fourth Edition

Arriba Publishing, Colchester, Vermont

    Printings: 1987, 1989, 1991 (SecondEdition),
    1994 (Third Edition), 1998 (Fourth Edition), 2002

Grateful acknowledgment is given for permission to use
material from *The Feeling Good Handbook* by David D. Burns,
M.D.. Copyright © by David D. Burns, M.D.. Reprinted by
permission of William Morrow & Co., Inc.

Library of Congress Catalog Card Number: 86-90583

Published by: Arriba Publishing
                 2238 Middle Road
                 Colchester, Vermont 05446
                 Phone: 1-800-639-4178

Printed in the United States of America

Publisher's Cataloging-in-Publication Data.
Murphy, Bill
  Greener Pastures on Your Side of the Fence
  Better Farming with Voisin Management Intensive
  Grazing
  Includes index.
1. Pasture Management - Handbooks, manuals
2. Grazing Management - Handbooks, manuals
3. Livestock Production - Handbooks, manuals

ISBN 0-9617807-3-8 Softcover

To Lita

¡Nos hemos divertido harto!
¡Y seguimos pasándolo bien, mejor que nunca!
El amor entre nosotros crece dia a dia.
¡Pensar que todo empezó con una mirada por un agujero!

Yo te queria siempre,
Antes que te vi,
Antes que me miraste
Por el agujero.

Cuando te vi, te reconoci´
Mi pasado,
Mi futuro,
Mi eternidad.

Cuando miro por los agujeros de tus ojos negros,
Veo el gran amor que tienes para mi,
El gran amor que eres:
El pulso del universo.

Acknowledgments

This book resulted from all of my experiences up till now. So everyone who has touched my life in one way or another has contributed to it. In particular, Joe and Dolores Murphy, who raised me on a farm and helped me develop a farmer's perception, sensibilities, and common sense. Jesse Scholl, who got me started doing pasture management research. And my friends Lita Young, Oscar Mena Barreto, Rick Donahoe, and Alan Hamerstrom, who opened my eyes in more ways than they know. I'm grateful to you all!

Caroline Alves, Doug Flack, Alice Pell, Dan Patenaude, Jeanne Murphy Patenaude, and Henry Swayze reviewed the original manuscript, asking questions, and suggesting changes. Through the years Dan, Jeanne, Doug, and Henry have continued to mutually learn with me how to manage our pastures better. And Murray Thompson, you keep asking hard questions that force me to get it right! Thank you all very much!

During my early stages of doing pasture management research, two people made it happen. David Dugdale worked with and helped me for several years at the University of Vermont. John Rice was the first Extension Agent in Vermont to see the value of feeding livestock on well-managed pasture instead of unnecessarily in confinement. John risked his reputation in 1984 by persuading three dairy farmers to manage the grazing of their cows and participate in a study to measure pasture forage quality and yield. Thanks again to you both!

Mike and Tammy Hanson gave me the opportunity to manage the grazing of high-producing cows using their Holstein herd in 1992. This experience reassured me that this is the way to farm. Thank you for trusting me!

Allan Nation deserves a lot of credit for helping to develop grazing management expertise by sponsoring conferences and tours, and facilitating a valuable international grazier network with his publication, *The Stockman Grass Farmer*. He is leading the way where others have feared to tread or trample. Way to go, Allan!

Allan Savory's Holistic Management concepts have made it easier to consider social, environmental, and quality-of-life issues related to production and economics that scientists and politicians have ignored in the past to the detriment and peril of us all. Thank you, Allan!

Many thanks to Tim Barrows, Diane Bothfeld, Jack & Brent Brigham, Ian Brookes, Bill Bryan, Louise Calderwood, Emily Carlson, Jeff Carter, Annie Claghorn, John & Judy Clark, John Cornell, Dan Dindal, Mike & Barbara Eastman, Nancy Everhart, Sarah Flack, Henry, Sally, Travis, & Amy Forgues, Catlin Fox, Stew Gibson, Willie Gibson, Dave Hoke, Justin Johnson, Dennis Kauppila, Marjorie Major, Lisa McCrory, Alexandre Mena Barreto, Elena Metcalf, Chet Parsons, Brian Pillsbury, Ed Rayburn, Mark & Sarah Russell, Jon & Beverly Rutter, Paul & Doris Seiler, Joshua Silman, Abdon Schmitt, Bill Snow, Jim Welch, Jon Winsten, Ted Yandow, Peter Young and all of the more than 120 farmers who participated in our Grazier Support Network. You truly are making a difference!

We farmers, economists, Extension agents, editors, feed consultants, veterinarians, writers, seed and grazing equipment companies, students, pasture management consultants, and researchers are mutually learning how to increase pasture and livestock productivity, to improve farm profitability and family quality of life and maintain our rural culture. Together we can do it!

Contents

Preface

In the dark forest a berry drops:
The sound of water.

Basho

The above words create a beautiful scene complete with sound in the mind's eye. On another level they describe exactly how I felt about this book when I first wrote and published it several years ago. This book was like that berry dropping into the large pond of agricultural writings and practices. In this book I describe a method that I know can help farmers to farm more naturally, efficiently, and profitably, and improve their quality of life. The ripples from this book and others are spreading, but most farmers still are unaware of the method. Permanent pastures in humid-temperate regions of North America generally still are a neglected resource, producing far below their potential the way they usually are managed. Clear guidelines such as those presented here are needed for proper management of pastures in these regions.

Vague management suggestions, such as "Rotational grazing involves grazing each of a series of pastures in rotation and then moving animals to the next." or "Rotation every 6 to 10 days of three or more pastures will allow the plants to recover." actually can do more harm than good. Following these kinds of general statements, farmers divide pastures into two or three parcels and pride themselves on it, only to have their animals run out of pasture forage in July or August. The pastures then get blamed for producing so little. But it isn't the pastures' fault! Low livestock productivity on pasture usually is due to extremely poor management of a forage crop in a pasture situation.

Understandably, many people in the United States question the value of rotational grazing. This attitude partly reflects conclusions about past research results that showed only an 8- to 10-percent gain in livestock productivity from rotational grazing, compared to continuous grazing. Researchers thought that this gain was too small a return for the extra effort and expense, so they recommended against using rotational grazing.

But if looked at in light of what we know now, it's clear that the rotational grazing designs that researchers tried didn't adequately meet plant or animal needs, and therefore didn't really differ enough from continuous grazing in their effects on pasture plants and grazing livestock. The designs all had at least three basic defects that prevented their success:

• They didn't take into account the need to vary the recovery period between grazings as plant growth varies due to changes in growing conditions.
• They didn't sufficiently limit the period that animals grazed a paddock at any one time.
• They didn't provide livestock with plants at a growth stage that enabled high intake of high-quality forage.

The problem of pasture management in the United States has been dealt with in essentially two ways. One way has been through pasture renovation. This is an expensive process in which the sod is killed through use of herbicides or repeated cultivations, lime and fertilizer are applied, the seedbed is prepared or a sod-seeder is used, and a mixture of selected legumes and grasses are seeded. In short, an attempt is made to deal with permanent pastures as if they are field crops. But except for machine harvesting surplus forage, most field-crop technology can't be transferred to permanent pastures, even with major modifications. A permanent pasture is not the same kind of ecological environment as a field crop.

The final instruction in this renovation process has been: "Remember to change the grazing management or the pasture will become the same mess that you had before renovating." But no one stated exactly what that management change should be, so of course pastures didn't remain improved for very long.

Renovation of permanent pastures on much of the land in North America is economically or physically impossible anyway, because of steep slopes, shallow, rocky soils, ledge outcroppings, boulders, and brush. The land is considered to be marginal and is in permanent pasture because it is unsuitable for tilled field crops. Besides, good evidence indicates that renovation isn't necessary. Grazing management of the pasture must be changed first, not last. Nothing else may be needed.

The other way that people have dealt with the problem of pasture messes has been to avoid pasturing or pasture management, through year-round feeding of machine-harvested and stored forage to livestock in confinement. Some confinement farmers continue to place animals on pasture, but the pastures generally are considered to be mere holding or exercise areas that have little or no feeding value. Stored forage and supplements are fed year-round, in addition to feeding daily green-chopped forage during the growing season.

Although year-round confinement feeding did eliminate the need to manage pastures, it brought on many problems of its own, some much more serious than poorly managed pastures ever presented. Mastitis, leg and hoof problems, metabolic diseases, breeding difficulties, and high culling rates are much more prevalent in confined animals than among animals that graze during most of the year. All of these are symptoms that something is wrong, and all increase production costs, which decreases profitability.

The biggest problems with confinement feeding is that it requires high capital investment in facilities and equipment, large amounts of purchased supplements, and excessive labor demand. Feeding in confinement costs two to six times more than when animals graze their forage.

Unlike other businesses, farmers usually can't pass high production costs on to consumers. Farmers, especially in dairying, are caught in a cost-price squeeze situation, in which higher operating costs together with lower prices for their products have resulted in greatly reduced profit margins.

Supposedly the large investment was made to save labor and increase profits, but it has turned out that farmers feeding year-round in confinement work *a lot* more, have lower profit margins, and consequently have a poor quality of life compared to farmers who feed their livestock on well managed pastures during most of the year. Having grown up on a dairy farm, it's hard for me to understand why any sane person would choose to extend winter confinement feeding year round. We always counted the days until those cows could be turned out on pasture in the spring!

All in all, an end result of the shift to year-round confinement feeding is that pastures have been practically eliminated from American farming experience for a generation. It's just as well, though, because pastures never were properly managed in this country, and the old ways are better forgotten. Also, new electric fence chargers and fencing materials developed mainly in New Zealand are available now to make a return to pasturing in the United States relatively easy.

New Zealand's highly productive and profitable agriculture depends almost entirely on permanent pastures, using grazing management initially defined by Andre Voisin of Normandy, France. Voisin taught biochemistry at the National Veterinary School of France and at the Institute of Tropical Veterinary Medicine in Paris. He also was a Laureate Member of France's Academy of Agriculture, and held an Honorary Doctorate degree from the University of Bonn, Germany. Voisin also farmed in Normandy, and he remained essentially a farmer in his outlook. That is, he was able to observe and understand natural interrelationships that often are missed by people not as perceptive and in tune with nature as farmers are.

Voisin saw pastures differently than other people did. He felt that if we take care of the pasture plants and soil life, they will take care of the grazing animals. And that is just what happens. For example, New Zealand farmers feed almost the same number of cattle as there are dairy cows in the United States, and five to seven times more sheep (plus 1 million each of deer and goats), but they do it on a pasture area the size of Wisconsin! And without grain supplements!

In the United States some people refer to the grazing management method used in New Zealand as intensive rotational grazing, but this may be misleading, and there's a lot more to it than what that may imply. Voisin called his flexible grazing method "rational grazing" to emphasize the logical, thoughtful management based on observation, needed in rationing out pasture forage according to the needs of livestock (just as feed is rationed out in confinement feeding), while protecting the plants from overgrazing and achieving a high level of forage use.

Other terms being used to describe this grazing method include short duration, controlled, planned, and intensive grazing management. Short duration is confusing, since it emphasizes grazing period in a paddock, when actually recovery time between grazings is more important. Intensive grazing implies grazing plants to a short residual, which might be confusing because taller residues sometimes are okay. Voisin mentioned controlled grazing in his book, but preferred to use a new term to describe his definition of basic elements of the method in a way that he felt would be useful to farmers.

Recently Jim Gerrish of the Forage Research Center in Missouri suggested that "management intensive grazing" more accurately describes the method, by emphasizing that management is the most important part of grazing. I think that this term may be a better translation for our purposes of what Voisin meant in French, rather than the literal translation that has been used. Here I'll use "management intensive grazing" for the method used under humid-temperate conditions.

If you're grazing livestock on rangeland in a brittle environment, you need to use planned grazing management. This is Allan Savory's term to describe the method when applied to rangeland, where management is complicated by drought, soil crusting, wildlife requirements, and public use of the land. It involves planning for specific recovery periods between grazings, monitoring regrowth of individual severely grazed plants, using herd effect to break soil crust to allow water penetration and seedling establishment, wildlife reproduction and feed needs, and hunting seasons. All environments fall somewhere on a continuous scale from brittle to nonbrittle. Unreliable precipitation regardless of volume, and poor distribution of atmospheric moisture through the year characterize brittle environments. For information contact: Center for Holistic Management, 1010 Tijeras NW, Albuquerque, NM 87102; Phone: 505/842-5252, Fax: 505/843-7900, e-mail: chrm@igc.apc.org.

Whatever you call the method, make certain that what you do in practice includes the key elements of adequate plant recovery time between grazings, correct forage allocation for livestock, and short grazing periods with high stocking density. Whether you're grazing pasture or rangeland, planning and monitoring certainly must be part of your management.

Management intensive grazing can be used in North America just as well as it is in New Zealand and elsewhere, to allow pastures to reach their full productive potential. The principles are the same, but adjustments are needed to fit local conditions.

What I present here may seem to be slanted toward permanent pastures, rather than pastures that are part of a crop rotation (rotational pastures), but this is only because most farmers right now want to increase the productivity of their so-called marginal land, which usually is in permanent pasture. Also, this "marginal" land is ideally suited for ruminant livestock production. Ruminants can use forage grown on such land to produce meat, milk, and wool, and not compete with production of other human food and fiber on arable land. Everything in this book applies to rotational pastures, but differences certainly exist. For example, earthworm numbers in the soil and plant density and rate of regrowth of a pasture rotated with tilled field crops will be much less than those in an adjacent permanent pasture, even though both are intensively managed. I point out important differences if they relate to management, so that they can be taken into account when dealing with rotational pastures.

As more and more farmers return to proper 5- to 7-year crop rotations (as they surely must do before monoculturing or two-crop sequences destroy the soil and ruin the environment), pastures will become essential parts of crop rotations again. I hesitate to call these kinds of pastures rotational pastures, because it results in such terrific confusion of terms, and implies that these pastures are being rotationally grazed, when actually they might be continuously grazed. So why don't we now, before pastures commonly become part of crop rotations again, agree to use the sensible British term, "leys", for such pastures? In this book, at least, leys are pastures that are part of tilled crop rotations.

It's unlikely that prices farmers receive for their products will increase (Prices probably will decrease as free-trade agreements go into effect and government subsidies end.), but you can improve your profit margin now, by reducing production costs. This is a book for farmers who want to improve the profitability of their farms and reduce their work load, while properly and responsibly caring for the land and the plants and animals that live on it. It isn't evidence in support of a method that I want you to believe in. Believing is not knowing!

This book explains a technique enough so that you can try it and experience it for yourself. Then you will *know* that it works!

Probably the greatest benefit arising from using pastures as valuable parts of farms, isn't in terms of dollar profits, but the peace of mind and mellowness that naturally develop in farmers as they get back in touch with the land, and stop running to feed and clean up after confined livestock. It's a lot better to just let the livestock go to the feed and spread their manure themselves.

Pastures are fun! So shut off that 160-horsepower articulated tractor, climb down from that air-conditioned, stereo-rocked cab, and take a leisurely walk around your pasture. Lie down under an old tree in the pasture. Chew on some clover flowers. Enjoy this wonderful world we live in!

1
Pasture Plants

If I look closely I can see
The shepherdspurse
Blooming beneath the hedge.
Anonymous

Permanent pastures are amazingly complex environments, compared to row crops such as corn. Twenty to
thirty plant species can be present, especially when the
pastures have not been grazed intensively. In a row-crop
situation one species, the crop, is present plus any weeds
that managed to escape being killed by tillage or herbicide
application. Even if the row crop is very weedy, still only
five to ten weed species usually are present.

Leys are actually more similar ecologically to tilled row
crops than to permanent pastures. Leys usually have only
two or three species each of legumes and grasses, plus a
few "weed" species, and are periodically plowed under for
row-cropping, only to be reseeded to pasture plants after a
few years. Overstocked, continuously grazed pastures,
whether permanent or ley, may only contain one legume
and one grass plus low-growing "weeds". These will be
plants that have or can assume a growth form close to the
soil surface below the animals' grazing height.

If you slowly walk through your pasture looking
carefully, you will find a world that you may not have
known existed. Take along field guides to grasses, legumes,
weeds, and wild flowers so that you can identify the plants
(see Recommended Reading list in Appendix). And take a
small shovel to see what is happening underground.
Usually in pastures we see the forest and not the trees.
Let's look at the trees!

As many as ten or more grass species may be present in
an undergrazed permanent pasture. Among others, in the
northeastern and northcentral United States these usually
include Kentucky bluegrass, quackgrass, timothy, Canada

bluegrass, bromegrass, orchardgrass, annual and perennial ryegrasses, tall fescue, and reed canarygrass. In extremely overgrazed pastures the only one present may be Kentucky bluegrass. All of these have different growth forms, growth habits, and carbohydrate reserve cycles.

Several legumes may be present, including alfalfa, red clover, white clover, alsike clover, birdsfoot trefoil, vetches, and sweetclover. In overstocked, continuously grazed pastures, white clover may be the only legume.

In low-lying areas of pastures, reed canarygrass, sedges, and/or rushes grow, depending on how wet the areas are. Sedges and rushes have feed value, but aren't very palatable to animals. A beneficial thing to do with areas growing sedges or rushes is to fence them off from the rest of the pasture, so that livestock can't walk through them, and they'll become refuges for birds and other wildlife.

A large number of broadleaf plants (forbs), such as plantain, dandelion, chicory, burdock, and milkweed also are present in pastures. The number and diversity depend on grazing management. More species in greater numbers will be present in well-managed or understocked pastures. Fewer species and only low-growing ones, except for those that protect themselves from grazing (e.g. thistles), can survive in overstocked, continuously grazed pastures.

Another group of plants that you'll find include some of the oldest plants in existence: ferns, horsetails, and lichens. They are found especially along edges of woods in moist or shaded areas. These plants generally are good feed for livestock, except for braken fern and horsetail, which are poisonous.

When you start managing grazing be careful, so that you don't force your livestock to eat poisonous or other harmful plants that they normally would not eat. It isn't that they can't eat any poisonous plants; a few might not harm them. The problem develops when animals eat large amounts of poisonous plants in proportion to nonpoisonous ones. To be safe, either remove harmful plants or exclude areas containing them from grazing (see chapter 2 and Appendix).

Finally, the most obvious plants you'll see, perhaps more than you may want to, are bushes, shrubs, and trees. These can cover large areas of a pasture, especially when it has been understocked and continuously grazed. Bushes and shrubs seen from the side at our sight level may seem to be more dense than they actually are. Scattered bushes, shrubs, and trees actually block very little sunlight, help hold the soil in place, recycle nutrients that have been leached below pasture plant rooting depth, and provide browse and shade for livestock; some fix nitrogen.

Check to see if the trees, bushes, and shrubs are far enough apart to allow sunlight through for grass and legume growth. If they're scattered enough, don't worry any more about them because they won't increase under management intensive grazing. If they're growing too densely, thin them out until they are more or less equidistantly spaced (at least 30 feet apart in all directions) and cover about 30 percent of the area. Leave as many trees as possible, because the Earth needs all the trees that can be grown, and scattered trees are desirable for shade. Thinning can be done with mob stocking (see chapter 10) or with a chain saw.

Nature always moves toward the most well-adapted and stable group of plants in any situation (climax vegetation). For example, bushes and shrubs are an intermediate stage in the movement toward a tree climax in northeastern and northcentral United States. This region used to be covered by forests, and growing-condition forces move it rapidly back to that stage, unless work is done to maintain it as grassland. Managing livestock grazing is a way of doing that work while reaping the benefits from highly productive pastureland.

Understocked, continuous grazing actually helps to move pastures along to a climax vegetation. Animals select the most palatable plants, usually leaving tall-growing broadleafs, bushes and shrubs, and young trees. These then have less competition and can grow more quickly, and the climax stage is reached sooner, unless someone intervenes with saws and land-clearing equipment to start the cycle all over again. By managing livestock grazing, you avoid this problem and expense.

If this walk through your understocked pasture occurs in mid to late summer, you will be greeted by a profusion of colorful flowers. People with highly developed sensibilities have said that colors of plants and flowers reflect cosmic forces. According to them, blue flowers reflect the influence of Saturn, Jupiter brings forth white or yellow flowers, Mars can be recognized in red flowers, and in green leaves we see essentially the Sun. In a pasture, certainly there is more than meets the eye!

In a way, it's a shame to manage a pasture more intensively, because the plants flower less under such management. But since plants reflect cosmic forces, a well-managed pasture with its almost throbbing green color, only indicates that more Sun is being expressed than other influences. This is important to your farm because solar energy drives everything on it (see chapter 2).

Along with the decrease in flowering that occurs, the pasture plant community simplifies, as species that can't withstand close grazing disappear. Always manage grazing to increase plant species diversity as much as possible, while maintaining high plant and animal productivity levels. Let's look more closely at plants likely to persist under management intensive grazing.

GRASSES
Growth forms are one of the obvious things that affect how grasses relate to other plants in a pasture and respond to grazing management. You will find grasses with two main growth forms in your pastures:
• Bunchgrasses (e.g. orchardgrass, perennial ryegrass, tall fescue, timothy) grow in clumps or bunches in the spot

where the original seed germinated and established. Open space exists between bunches that can be filled by other plants, such as legumes and sod-forming grasses.

Under consistent close (allowed to grow 4-8 inches tall, then grazed down to about 1 inch tall), well-managed grazing the upright clumpiness of bunchgrasses can be changed to a more prostrate growth form that produces abundant leafy forage.

• Sod-forming grasses (e.g. Kentucky bluegrass, smooth bromegrass, reed canarygrass, quackgrass) spread away from and around the spot where the original plant established. Many of these grasses have underground stems (rhizomes) that produce roots and new shoots at various distances away from the mother plant. Others have similar stems lying on the soil surface (stolons) that produce shoots and roots at nodes contacting favorable soil conditions. The new plants in both kinds of sod-formers eventually become independent of the mother plants, and produce new plants of their own. The rhizomes and stolons, along with the extremely finely divided, dense, moderately deep root systems common to most grasses, make these grasses especially valuable for soil and water conservation.

Most sod-formers have abundant growing points (terminal meristems) and leaves near the soil surface, so they tolerate and produce well under close, well-managed grazing.

Growth habit is another obvious characteristic of plants that affects how they respond to grazing management:

• Low-growing grasses (e.g. Kentucky bluegrass, perennial ryegrass) are well adapted to close grazing. They produce many side shoots (tillers) from their bases early in the season. Their growing points remain close to the soil surface below grazing level, and a thick, vigorous leafy sward can be maintained under close, well-managed grazing.

- Intermediate-growing grasses (e.g. orchardgrass, timothy, quackgrass), although similar in height, respond differently to grazing management mainly because of differences in locations of growing points and ways they store carbohydrate reserves. For example, orchardgrass and timothy persist and produce high quality forage under consistent (throughout each grazing season, year after year) close grazing that begins early in spring, but quackgrass appears to require lax grazing (allow to grow 10-16 inches tall, graze down to 4-6 inches tall) to persist. Under lax grazing, timothy also performs well, but orchardgrass forage quality declines.
- Tall-growing grasses (e.g. smooth bromegrass, reed canarygrass) apparently need lax grazing for best results. They may be damaged by close grazing, because their growing points become shaded by topgrowth and elongate enough so they can be removed by close grazing. Then regrowth has to come from buds on plant bases; these buds develop slowly. Depending on the grass and how much carbohydrate reserve it has, little or no regrowth may occur if growing points are grazed off.

  Stems of tall-growing grasses decrease forage quality, so producing animals must be given a large forage allowance (see chapter 5) so they can select more leaves. If you clip a tall-grass pasture, mow at 3 to 4 inches high to avoid cutting off growing points.

  Lax grazing of any pasture tends to result in thin swards with low white clover content. Annual overseeding of red clover or other tall-growing legumes and periodic nitrogen fertilizer applications may be needed to maintain high forage yield and quality.

Kentucky bluegrass (*Poa pratensis*) is an ideal pasture grass where it is adapted. It almost always exists in pastures because it can withstand even the worst grazing management, since its growing points remain very low to the ground. This same characteristic allows it to survive weekly mowings down to 1 inch in lawns. It is a strong sod-former that spreads by rhizomes.

Kentucky bluegrass has a bad reputation, because it doesn't perform very well under the continuous grazing that it usually is subjected to in the United States. Rundown, unproductive bluegrass pasture that stops growing in midsummer is due to extremely poor grazing management, not Kentucky bluegrass. (If alfalfa was treated the same way that farmers usually "manage" bluegrass, alfalfa would have as bad or worse reputation!)

Kentucky bluegrass produces most during cool parts of the season and grows slower in midsummer. When managed well, it produces high yields of excellent quality forage, and grows more uniformly throughout the season.

Kentucky bluegrass combines very easily with other grasses and legumes, forming mixtures that are well adapted to management intensive grazing. When you overseed established pasture or seed down a previously tilled-crop area, request seed of a pasture-type variety, rather than common bluegrass, which might be a turf-type that may have characteristics contrary to what you need.

Canada bluegrass (*Poa compressa*) is less desirable because it produces less forage that is stemmier, and it begins growth and flowers earlier in the season than Kentucky bluegrass. (I think many people confuse Canada bluegrass with Kentucky bluegrass, contributing to Kentucky bluegrass's undeserved bad reputation!) The flowering stems are very unpalatable, especially to cattle, possibly because they are unable to bite through the stems. Since ruminants only have front teeth on the bottom jaw, they eat by grasping plants with their tongues, teeth and upper mouth pad and tearing the plants off. When they do this with the flowering stems of Canada bluegrass and other plants, the stems probably cut their gums. So cattle may avoid eating plants such as Canada bluegrass and any neighboring plants for fear of getting bluegrass stems between their teeth. By beginning grazing early in the spring, flowering stems can be grazed off before they elongate very much and become too fibrous to be torn off. If it flowers before it can be grazed, mowing paddocks a few hours before animals leave each paddock, will allow animals to eat the stems and other plants growing among Canada bluegrass.

Canada bluegrass tends to grow better than Kentucky bluegrass on soils having lower levels of nitrogen, phosphorus, potassium, calcium, and magnesium. So as you improve the soil's fertility and seed Kentucky bluegrass, it will replace Canada bluegrass.

Perennial ryegrass (*Lolium perenne*) is another ideal pasture grass where it is adapted. When grown under high soil fertility (especially nitrogen), good moisture conditions, and mild winters it produces large amounts of excellent quality forage. The high quality of ryegrass forage results from the plants accumulating large amounts of nonstructural carbohydrates (stored energy) during spring and fall. These concentrations can reach 30 to 40 percent of the plant's total dry matter. Ryegrass has a ratio of digestible to indigestible fiber of 4:1, compared to 1:1 or less in orchardgrass. This readily available energy can support high levels of plant growth or livestock production. Dry matter digestibility of ryegrass forage declines as it matures, but not as much as occurs in orchardgrass. Where well adapted, ryegrass is very productive, combining well with white clover.

Grazing management is extremely important in maintaining ryegrass, especially in areas with cold winters. Perennial ryegrass is a prime example of how close grazing results in lateral spreading and rooting of tillers to produce a very dense, leafy sward. It needs to be kept grazed short in spring so that it tillers and thickens the stands. If not kept short, ryegrass tillers are killed by shading.

If you live in growing zone 4 or colder, you probably won't find ryegrass in your pasture. Late-maturing, cold-tolerant varieties are available, however, that might be useful to you. In this zone don't graze ryegrass shorter than 2 to 3 inches from the soil surface beginning about October 1, allowing it to grow to 5 inches tall before winter, so that it has reserves to live on over winter and forage mass to catch snow for insulation.

Don't despair if you can't grow ryegrass in your area because of cold conditions. The plants that you can grow, such as Kentucky bluegrass, orchardgrass, bromegrass, timothy, and quackgrass produce as much or more forage than ryegrass, unless you're willing to apply nitrogen fertilizer at more than 110 lb/acre/year.

Several years ago I began an experiment to introduce new grasses into my old Kentucky bluegrass-white clover sward. A Dutch plant breeder who had sent me 22 ryegrass selections came to see how the experiment was going. When he saw my vigorously growing dense sward (into which almost no ryegrass had established) he said, "Why are you trying to grow perennial ryegrass, when you can grow Kentucky bluegrass like this? Bluegrass is a better plant!" He then asked if he could collect bluegrass plants to take back to Holland for study and multiplication. He said that he had never seen such vigorous bluegrass plants, and felt that my plants were evidence that Kentucky bluegrass had originated in North America! Needless to say, after that I stopped trying to substitute perennial ryegrass for Kentucky bluegrass.

Orchardgrass (*Dactylis glomerata*) is a bunchgrass with an intermediate growth habit. Orchardgrass produces tillers throughout spring and leaves that have been grazed continue to elongate, so it recovers quickly from grazing. It needs to be grazed closely to keep shoots palatable to livestock; otherwise it tends to become coarse, tall, and very bunched; at that stage it is not readily eaten.

Orchardgrass is a valuable pasture plant that combines easily with legumes, and is well adapted to moderate soil fertility and low soil moisture conditions. It begins growth earlier in the season than associated legumes, so pasture containing orchardgrass should be grazed early and frequently to keep it in a young leafy stage, and prevent it from shading out legumes.

Recently developed varieties mature later, distribute seasonal growth more uniformly, are more disease resistant and yield more forage of higher feeding value than common orchardgrass. Some new varieties are fine-leaved and have a prostrate, many-tillered growth habit that produces a denser sward than common orchardgrass. Some varieties may tolerate drought better than others. Including early- and late-maturing orchardgrasses in different paddocks can make early spring management easier, because not all paddocks would need to be grazed at the same time.

Timothy (*Phleum pratense*) is an intermediate-height bunchgrass that performs well under close or lax grazing. It stores carbohydrates in a swollen lower area (corm) of the shoot for use in producing regrowth. You can see or feel the corms by looking at or touching the plants just above the soil surface. Timothy probably is the best all-around grass for palatability, persistence, establishment ease, yield, disease resistance, and tolerance of poor drainage. It tolerates almost everything but drought.

Timothy competes very little with associated legumes, so it can add forage yield and ground cover to mixtures without affecting legume production or persistence. Although timothy doesn't produce a high forage yield, it improves palatability of the overall sward.

Smooth bromegrass (*Bromus inermis*) is a tall-growing grass that forms a dense sod as it spreads throughout pastures by short rhizomes. It is extremely winter hardy and tolerates hot, dry conditions. Bromegrass provides excellent early spring grazing, when its growth is mostly leaves. If you don't hold bromegrass down by consistent close grazing, it will grow upright and its growing points will elongate in mid- to late-May. Then you'll have to leave a long residue (4 to 6 inches) in paddocks that contain bromegrass. You'll also have to allow bromegrass to reach a tall pregrazing height of 10 to 16 inches so it can accumulate carbohydrate reserves and persist.

Bromegrass combines very well with alfalfa for both hay and pasture, forming a particularly useful mixture for drier soils and droughty conditions. It also combines well with red clover and birdsfoot trefoil, but can shade out white clover unless kept short. New varieties of bromegrass selected mainly for pasture can persist longer and produce more forage than strictly hay types.

Bromegrass relatives such as Matua prairie grass (*Bromus wildenowii*), Gala grazing brome (*Bromus stamineus*), or Hakari mountain brome (*Bromus sitchensis*) probably won't be found in your pasture, but you might consider seeding them. They were developed in New Zealand to provide high forage yields during cool conditions of fall and winter, and dry periods of summer and early autumn. They grow well on drier sandy soils where other grasses have difficulty persisting. They can outproduce perennial ryegrass and orchardgrass, and their forage is highly palatable with excellent feeding value at all growth stages.

They can withstand close grazing of less than 4-day grazing periods, but they must be allowed a recovery period of 25 to 45 days until they again reach the seed boot stage. Grazing needs to begin when plants reach 10 inches tall and stop when they are 3 to 5 inches tall. Plants die if stockpiled in autumn or allowed to lodge anytime. Close grazing late in fall improves their winter survival and spring regrowth. If their sward density begins to decrease, allowing a long recovery period in summer results in reseeding and increasing plant numbers.

Reed canarygrass (*Phalaris arundinacea*) usually grows in fertile, low, moist areas of your pasture. Reed canarygrass also grows well on upland soils, and is one of the highest yielding, drought-tolerant cool-season grasses. It is a tall-growing sod-former that spreads slowly by short, thick rhizomes and forms heavy tough sod that is often bunchy. It produces most of its growth during the cool parts of the season, beginning very early in the spring. It has to be grazed early and kept well grazed down to less than 12 inches, otherwise it becomes unpalatable to livestock.

Common reed canarygrass usually has poor palatability because it contains toxic alkaloids, but if livestock don't eat too much of it, they aren't harmed. New varieties have been developed that contain less alkaloids and a higher leaf-to-stem ratio, making their forage more palatable and higher quality. Reed canarygrass combines well with upright-growing white clover, red clover, and alfalfa on upland soils.

Tall fescue (*Festuca arundinacea*) is an intermediate-height bunchgrass that also has short underground stems. If it is kept closely grazed or mown, it produces a dense sward and tight sod. Tall fescue is used widely for soil conservation purposes, turf, and forage. It tolerates continuous close grazing, and is particularly valuable for grazing beef cattle in spring, fall, and winter. If cattle are removed in midsummer and forage is allowed to accumulate until the first frosts occur, tall fescue pastures can carry cattle through the winter. Autumn frosts improve its palatability. New varieties are available that are free of endophyte and much more palatable than old varieties.

Meadow fescue (*Festuca pratensis*) is shorter and narrower leaved than tall fescue. It grows in areas where timothy grows, and is especially useful on wet soils. Meadow fescue performs well under close or lax grazing.

Quackgrass (*Elytrigia repens*) is another grass that you will almost always have in your pasture. In temperate areas throughout the world, this sod-forming grass probably holds more soil in place than any other plant! Although it can be a serious weed in tilled crops and gardens, it is a valuable grass to have in a pasture.

Immature quackgrass forage quality equals that of timothy and smooth bromegrass, and it can yield more forage than orchardgrass. It can contribute significantly to forage production, especially early in spring and under droughty conditions of midsummer.

Quackgrass can compete strongly with other plants, especially legumes, but if grazing is managed well this is not a problem. Quackgrass seems to require lax grazing with a tall (10-16 inches) pregrazing height and a long residue (at least 3 inches) to persist and produce well.

Warm-season grasses such as big bluestem (*Andropogon gerardii*), indiangrass (*Sorghastrum nutans*), and switchgrass (*Panicum virgatum*) may be in your pasture, depending on location. These tall-growing grasses have short rhizomes that tend to produce a bunch-like appearance and coarse sod. They are valuable wherever management and conditions enable them to grow (see chapter 9).

LEGUMES
You should find at least scattered legumes in your pasture, no matter how poorly it has been managed, even though animals selectively graze legumes. If you don't find any legumes, a soil fertility or pH problem most likely exists. It's absolutely essential to have 30 to 50% legume content in your pasture to obtain the excellent quality forage needed to achieve high livestock production levels at low

cost. Compared to grasses, legumes contain more protein, double the mineral content, less fiber, and have a higher ratio of soluble to insoluble carbohydrates.

Although air contains 80 percent nitrogen, it is unavailable to most organisms. Legumes don't require soil nitrogen because they can use nitrogen from the air. Rhizobia bacteria that associate symbiotically with legumes in nodules on the plant roots, combine (fix) atmospheric nitrogen so that it can be used by the legumes and, consequently, by all other plants and animals. This nitrogen ultimately becomes available to associated grasses through the urine and manure excreted by grazing livestock, and decomposition of legume nodules, roots, and shoots.

Because legumes are so valuable to you, make notes of where legume sward content is less than 30 percent during the grazing season, and sod-seed, overseed, or frostseed (see chapter 10) the legumes you want every spring where needed. When seeding legumes always inoculate the seed with fresh rhizobia inoculant that's for the legume specie you are using, even if the seed has been preinoculated. Selected inoculant strains of rhizobia can fix more nitrogen than native rhizobia in the soil.

White clover (*Trifolium repens*) usually will be the main legume sustaining humid-temperate pasture productivity. It withstands the most intensive grazing, fixing large amounts of nitrogen while producing high yields of nutritious forage.

White clover grows rapidly, spreading by creeping stolons lying on the soil surface. Kneel down and follow a white clover stolon with your fingers. Notice how every few inches roots have formed at nodes. These rooted spots eventually become independent plants that will then spread on their own. Carefully dig up one of the larger rooted plants and separate the soil from the roots. On the very finest root hairs you should be able to see pinkish white, round formations. These are the nodules in which nitrogen from the air is being fixed by rhizobia, making it available to all other living organisms, including yourself.

While you're kneeling down, carefully dig and loosen more clover roots and some grass roots from the soil. Notice how intimately roots associate with the soil. With an electron microscope you would be able to see that life is continuous between plants and surrounding soil. No distinct line exists between life within plants and soil life.

Since it grows so low to the ground, white clover withstands grazing very well. No matter how closely grazed, some leaves always remain below the level of grazing. This enables white clover to continue intercepting solar energy and converting it through photosynthesis to carbohydrates and plant regrowth. The white clover forage that animals eat is all leaves, petioles, and flowers; so the forage has extremely high feeding value all season.

White clover grows on a wide range of soil conditions, including acid soils down to about pH 4.5, as long as the soil contains enough calcium and other needed nutrients. White clover produces best on fertile, moist soils, and doesn't do well on light sandy soil because of its shallow root system. It grows best at a soil pH of 5.5 to 6.0, because the rhizobia that infect clovers and fix nitrogen function well in that pH range. Healthy looking white clover plants with pink nodules indicate that soil conditions are favorable or can be improved easily.

When seeding white clover, use selected varieties now available that can fix five times more nitrogen than common or "native" types. (Note: New Zealand common white clover seed probably is Huia, so you don't need to pay for certified Huia.)

Red clover (*Trifolium pratense*) is more deeply rooted than white clover and, consequently, can grow under drier soil conditions. Otherwise it grows well under soil conditions similar to those favored by white clover. Because red clover grows more upright than white clover, it can't survive close grazing as well as white clover can. Red clover will persist, however, especially in areas that you set aside in the spring for machine harvesting.

Because of its upright growth habit, red clover combines well with tall-growing grasses, and forms very productive mixtures useful for haylage, hay, and pasture. Red clover plants usually only live for 2 years, so you should overseed it each spring where needed. By doing this with disease-resistant varieties you can maintain a high content of this valuable legume in swards.

Alfalfa (*Medicago sativa*) is the most deeply rooted legume that we use, and it can grow on the driest soils. In fact, it doesn't do well at all on soils that are too moist or poorly drained. Alfalfa is a very productive legume under conditions where it is well adapted and well managed. Because the rhizobia that infects alfalfa requires a soil pH of 6.0 to 7, you probably won't find it in your pasture unless its soil pH is in that range.

Because of its upright growth habit and carbohydrate reserve cycle, alfalfa regrows mainly from stored carbohydrates, not from leaves left after grazing. So alfalfa usually can't withstand as frequent grazing as white clover, unless new alfalfa varieties selected specifically for grazing are used. Hay-type alfalfa varieties may not persist if grazed in the first and second spring rotations. At that time alfalfa's carbohydrate reserves are low, and it needs several weeks to replenish them through photosynthesis during the first growth of spring. Hay-type alfalfas can persist under management intensive grazing if the first alfalfa crop of spring is harvested for haylage or hay. By allowing the alfalfa to recover for 25 to 30 days after the first cutting, it can then be included in the grazing rotation without damaging the plants. This is because at that time recovery periods between grazings are long enough to meet alfalfa's needs.

Alfalfa combines well with most temperate grasses, and the mixtures produce large amounts of high quality forage. Alfalfa mixtures with bromegrass and reed canarygrass tolerate drought conditions, so they would be especially useful on dry, sandy soils where other grasses and white clover can't grow.

Birdsfoot trefoil (*Lotus corniculatus*) can grow in extremely adverse soil conditions. If your pasture soil is a clay that floods in spring and bakes as hard as concrete in summer, trefoil may do well there. It also grows well along roads where salt is spread during winter. While it will grow on poorly drained, acid, droughty, infertile soils, trefoil produces best on heavy fertile soils at pH 6.5. Besides fixing nitrogen and producing large amounts of highly nutritious forage, birdsfoot trefoil has the added advantage of not causing bloat.

All of this indicates that trefoil should be a great legume to have in your pasture. One problem is that trefoil has poor seedling vigor and establishment is slow. During establishment, competition from associated grasses and forbs must be decreased with proper grazing management. It is more difficult to establish and maintain trefoil in mixtures with tall-growing grasses (e.g. orchardgrass, bromegrass) than with Kentucky bluegrass, because the taller species tend to shade trefoil too much unless swards are kept short.

Trefoil maintains a low level of carbohydrate reserves during the growing season, regrowing mainly from solar energy captured by photosynthesis in existing leaves. This means that enough postgrazing trefoil residue should be left so that leaves remain to continue photosynthesizing for regrowth. Using prostrate, Empire-type varieties helps maintain a long residue with leaves. By keeping the grass-trefoil sward from growing more than 6 to 8 inches tall before grazing, you can prevent lower leaves on trefoil from being lost due to shading, thereby helping it recover quickly after grazing. Birdsfoot trefoil content in swards can be maintained or increased by allowing a long enough recovery period (about 6 weeks) in midsummer to enable trefoil plants to flower and set seed once every other year.

Other legumes such as alsike clover and vetches might be found on this leisurely walk through your pasture. Alsike resembles red clover in growth habit, and produces better than either red or white clovers on wet soils under cool conditions, and in lower soil pH levels. The vetches are viny legumes that climb up neighboring plants and persist in understocked pastures.

FORBS
Although this group of broadleaved plants generally is looked down upon by people and called "weeds", grazing animals have a different perspective. Many "weeds", such as plantain, dandelion, and chicory are actually preferred by grazing animals over the grasses and legumes that humans have selected. Almost all "weeds", except plants that protect themselves from being grazed with spines, thorns, odors, or poison (see Appendix), are edible and nutritious for livestock in early growth stages (see chapter 2). For example, dandelion forage that I had analyzed for feeding value contained 15% dry matter, 22% crude protein, 72% total digestible nutrients, and 0.76 Mcal NEL/lb DM!

To increase the diversity of plant species in your pasture, try overseeding Puna forage chicory in the spring. Because chicory is deep-rooted, it recycles leached soil nutrients and produces well during dry periods. Adding Puna chicory to grass-legume mixtures can increase livestock productivity. For example, it has been estimated that cows grazing mixtures with Puna chicory produce 2 to 3 lb/day/cow more milk than without Puna.

REFERENCES

Ballerstedt, P.J. 1993. An introduction to "Grasslands" Matua Praire Grass - a wonderful new grass, not a wonder grass. *The Stockman Grass Farmer.* 50(3):12-14.

Frame J. 1985. Do not scorn the secondary grasses. *Grass Farmer.* British Grassland Society. 23, 7-11.

Frame J. 1991. Herbage production and quality of a range of secondary grass species at five rates of fertilizer nitrogen application. *Grass and Forage Science*, 46, 139-151.

Barnes, R.F., and D.A. Miller, and C.J. Nelson (eds.) 1995. *Forages.* Iowa State University Press, Ames. 5th ed. Vol. I. 516 pp.

Jung, G.S. 1988. Use of ryegrass in the Northeast. U.S. Regional Pasture Research Laboratory, University Park, Pennsylvania. Mimeo.

Langer, R.H.M. 1973. Grass species and strains. *In*: R.H.M Langer (ed) *Pastures and Pasture Plants.* A.H. and A.W. Reed, Wellington, NZ.

Massey, S. 1993. What's new in grazier's gear. *Stockman Grass Farmer.* 50(1):14, 50(3):17, 50(9):20.

Nation, A. 1993. Production per man more important than per cow. *Stockman Grass Farmer.* 50(3):1-10.

Nation, A. 1994. Pasture species diversity stabilizes the forage curve. *Stockman Grass Farmer.* 51(9):1, 5-8, 19.

Nation, A. 1995. Quackgrass lacks respect but is excellent forage. *Stockman Grass Farmer.* 52(2):10.

Savory, A. 1988. *Holistic Resource Management.* Island Press, Washington, DC. 565 p.

Semple, A.T. 1970. *Grassland Improvement.* Leonard Hill Books, London. 400 p.

Smetham, M.L. 1973. Pasture legume species and strains. *In*: R.H.M. Langer (ed) *Pastures and Pasture Plants.* A.H. and A.W. Reed, Wellington, New Zealand.

Smith, B., P. Leung, and G. Love. 1986. *Intensive Grazing Management*. The Graziers Hui, Kamuela, Hawaii. 350 p.

Smith, D. 1981. *Forage Management in the North*. Kendall/Hunt Pub. Co., Dubuque, Iowa. 258 p.

Steiner, R. 1977. *Agriculture*. Rudolf Steiner Press, London.

2
Pasture Ecology

In the landscape of spring
There is neither high nor low;
The flowering branches grow naturally,
Some long, some short.

Zenrin

Changes in pasture plant populations occur because some individual plants are better adapted to their surrounding conditions and leave more descendants than others. This determines individual fitness. Aspects of individual fitness influence how pasture plants grow, reproduce, and spread.

Within a pasture plant community or sward, individual plants gain an advantage over their neighboring plants in various ways. The kinds of activities that give an advantage usually are selfish qualities that include taking up more nutrients than needed, transpiring more water than needed, and producing a denser canopy than needed to intercept an optimum amount of light. So a plant actually gains in fitness if its activity harms its neighbors, since fitness depends only on leaving more descendants.

Unfortunately, many of the things that favor individual plants of one species over another, or plants of one genetic makeup over another in leaving more descendants, work against the cooperative community aspects that matter to farmers. (Many smaller plant communities exist in every pasture sward, depending on several things including soil type, slope, aspect, and water-holding capacity.) The selfish qualities important to the individual plant may produce a pasture sward that's quite different from what's needed for good forage production. This is because the fate of an individual plant depends on its immediate neighbors, which because of their demands on environmental resources, directly influence its wellbeing. The processes involved concern fitness at the level of individual plants, not forage productivity measured at the community level.

To understand the nature of the forces that influence individual plants within a pasture sward, we need to consider what happens in the sward from the plant's point of view. If we know how plants experience their local environments, we might be able to influence that experience to favor improved forage production.

COMPETITION

As long as all necessary growth factors are adequate to meet the needs of all plants within a sward, competition doesn't occur, and no individual plants are favored over others. But as soon as the immediate supply of a single factor becomes less than the combined needs of the plants for that factor, competition begins. This is the point when individual plants begin to gain or lose in fitness, and changes begin to occur in the sward.

Competition among plants occurs for water, nutrients, light, carbon dioxide, oxygen, and means of pollination and seed dispersal. Other things such as temperature and humidity certainly affect plant growth and reproduction, but are not limited in supply, and therefore not strictly involved in competition. Except for some root crops, competition for space rarely occurs. Usually there is plenty of space for more plants, leaves, branches, and roots.

Most of the things that plants compete for exist as a pool, which individual plants draw from. If the pool of a certain growth factor is limited, the successful competitor is the plant that draws on the limited factor most rapidly from the pool, continues to draw on it when the supply is low, or is able to draw on it when other plants can't because the pool is too low. A successfully competing plant usually is one that quickly uses its immediate supplies of growth factors, and then extends into neighboring areas by growing more roots and leaves.

Almost all of the advantages gained by successfully competing plant species can be described by two words: amount and rate. These advantages include greater carbohydrate storage in seed or roots, more rapid and complete germination, earlier growth start in spring, faster growth of tops and roots, taller and more branching stems, deeper and more spreading roots, more tillers, more flowers, and larger leaves.

Any or all of these properties help individuals survive and leave more descendants. All of these traits express genetic and physiological backgrounds that enable the plants to take up water or oxygen from wet or dry soils, place leaves advantageously for light interception, take up nutrients that are less available, and adapt in growth and development to changes in climatic conditions.

GRAZING LIVESTOCK

In addition to all of the above influences, grazing livestock greatly affect the composition of pasture plant communities. Animals always cause a pasture to be a more complex mixture of plants than it otherwise would be. This is because animals graze selectively and in patches, and the effects vary in time and in space. Animals also drop their manure and urine in patches, which affects some plants more than others. Besides walking, running, and jumping on a pasture, animals sit, lie, scratch, and paw on it. All of these things result in a sward containing a wide variety of plants adapted enough to survive their different local conditions.

ENERGY

Available energy is what makes everything possible. Energy as we know it becomes available sooner or later from the Sun in the form of sunlight. This most expensive farm input, energy, is available "free" to those who understand that farming is essentially intercepting sunlight to produce saleable products (Figure 2-1).

Farmers make the most profit when they farm in ways that capture as much solar energy as possible and convert it to high-value products with the least amount of inputs. Permanent pasture-based farming does this best.

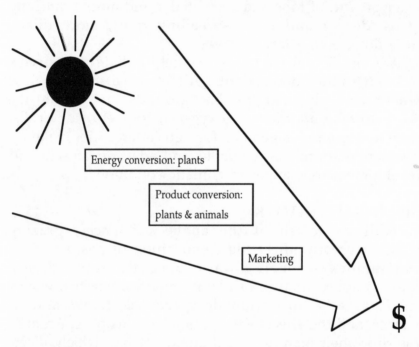

Figure 2-1. Solar energy chain from sun to solar dollars (Adapted from Savory, 1988).

The amount of solar energy intercepted can be increased by increasing the amount of leaves per unit area, size of leaves, and/or length of growing season (Figure 2-2). Increasing all three enormously increases solar energy interception and potential conversion to product. This is what makes pasture so much more potentially productive and profitable, compared to tilled annual row crops such as corn, which has low leaf area during much of its short growing season.

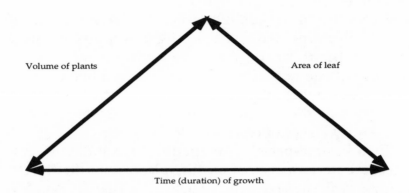

Figure 2-2. Surface level that controls solar energy flow (Adapted from Savory, 1988).

COMPETITION FOR SOLAR ENERGY (SUNLIGHT)

If it were possible to separate all the components of your pasture, sunlight certainly would rank as the most important and having the greatest influence on its botanical composition and yield. (Of course, nothing can really be separated from anything else: everything interrelates or "goeswith" everything else.) Now you would think that there would be plenty of sunlight for all the plants in a pasture, so why is there competition for it?

The reason is that sunlight decreases as it passes through a foliage canopy, because the leaves intercept and absorb it. If sunlight is reduced too much, growth and development of shaded plants become slower. The effect of shading depends on the plant's stage of development and how much it is shaded. Another characteristic of shading is that the reduced sunlight level in the canopy isn't the same over all plants in the canopy, or even all parts of single plants. Bottom leaves are shaded more than top leaves, and low-growing plants have more of their leaf surface shaded than tall-growing plants. Pasture plant canopies are surprisingly dense: a couple of inches of pasture canopy intercepts the same amount of sunlight as several feet of forest canopy!

Competition for sunlight occurs whenever a part of a plant interferes with the sunlight supply falling on another plant or part of the same plant, and the photosynthetic rate of the shaded leaf area decreases. The only times when there is little or no competition for sunlight are when plant growth first begins in spring and in the early stages of regrowth after grazing.

Two basic aspects of competition for sunlight greatly affect the performance of pasture plants, and are the main reasons why management intensive grazing works:

• Sunlight is not like a soil nutrient that remains in the soil until used. Sunlight must be used instantaneously as it passes through the canopy, or it is lost forever.

• The position of leaves within a pasture plant canopy is extremely important. Leaves that overtop other leaves gain a competitive advantage for sunlight.

Several properties of sunlight and plant canopy combine to influence competition for sunlight, and they change all the time. These include the sun's angle of elevation, whether the light radiation is direct or diffuse, density of the plant canopy, leaf angles, and light reflection, absorption, and transmission characteristics of leaves. So plants that persist in a pasture sward must contend with a daily and seasonally changing sunlight environment, especially during regrowth after grazing, plus withstand the effects of losing most of their leaf surface during grazing. It's amazing that any plants survive such treatment! Is it any wonder that pasture plants need longer recovery periods under adverse conditions and as the season progresses?

Effects Of Management
Plant height is extremely important in competition for sunlight. Even small height differences give considerable competitive advantages or disadvantages, especially

during leaf development or elongation. Anything that influences a plant's height or ability to shade in relation to its neighbors, determines its relative competitiveness. Pasture managers can influence three things that directly affect the relative competitive abilities of pasture plants for light: grazing frequency and intensity, and soil fertility.

Clover and grass contents of swards mainly reflect the variation in their abilities to compete for sunlight. For example, increased nitrogen from urine or fertilizer stimulates grass growth, which increases its ability to shade clover. More frequent grazing prevents shading of clover, so its competitive ability increases relative to grass.

Weedy pastures result from poor management that gives weeds a competitive advantage for sunlight. For example, weedy plants with long wide leaves suppress shorter plants such as white clover by shading, if a pasture is understocked and grazed infrequently. But this growth habit gives no competitive advantage under frequent close grazing with high stocking density. Allowing selective grazing of shorter desirable plants, gives uneaten tall-growing broadleaf weeds a great competitive advantage for sunlight that enables them to dominate the sward. Understocked lax grazing is the main reason for the weedy, unproductive messes typical of many North and South American pastures, because it gives weeds the competitive advantage for light over grasses and legumes.

WATER
Pastures usually depend on rainfall for water. When rainfall is irregular and soil water becomes deficient or excessive, plant productivity and persistence can be severely affected. Both drought and wet conditions can result in unexpected shortages of pasture forage.

Drought
One of the first consequences of dry conditions is that plant leaf area decreases. Because less solar energy can

then be intercepted, photosynthesis decreases, quickly resulting in reduced growth rate. Plants that can continue taking up water from drying soil immediately gain competitive advantage over plants unable to do so.

As conditions become drier, plants increase root growth to take up water from deeper in the soil, and restrict leaf growth even more to decrease water loss through evapotranspiration. Those that reduce their leaf area first or most, gain by decreasing water loss, but lose in competitive ability for sunlight. Plants that can increase their root system, without decreasing their leaf area, are the ones likely to dominate a pasture sward during and right after a dry spell.

The productivity and persistence of pasture plants under drought stress also depends on the levels and availability of soil nutrients before and during a dry period. When the water supply is limited, less nutrients can be taken up by plants. Much less nitrogen is fixed by legumes, probably because their rate of photosynthesis decreases, and nitrogen-fixing bacteria in the legume root nodules become starved for carbohydrates.

One reason why plant nutrition and water supply are so closely linked is that nutrient levels usually are highest near the soil surface, which is the first layer to dry out. Although plants may have roots in the deeper and wetter parts of soil, the low amount of nutrients in subsoil and unavailability of nutrients in dry surface soil, may limit plant growth and forage yield more than the lack of water itself.

So dry conditions adversely affect plants in two main ways:
• The shortage of water itself restricts leaf area.
• Nutrient deficiencies develop, further limiting leaf area.

Both directly affect plants' ability to compete for sunlight. For these reasons, pastures need especially good grazing management during and after dry spells to maintain a desirable plant composition in the sward.

Wet

When soil becomes too wet, plant growth slows. This occurs mainly due to water filling soil macropores and excluding oxygen needed by plant roots. The longer wet conditions persist, the more roots die of oxygen deprivation and the more fungal root disease organisms multiply and damage plants. Recovery can be slow after prolonged wet soil conditions; it takes time for roots to regrow and for plants to overcome fungal infections.

Always make certain that recovery periods at these times allow plants to regrow adequately before grazing again. (See chapter 5 for grazing management suggestions for dry and wet soil conditions.)

TEMPERATURE

As important as water is in a pasture plant community, temperature is even more important. The natural distribution of plants is determined by their adaptation to major climatic and soil aspects of the environment. Temperature limits the distribution and diversity of plants that can grow and develop in a region. As temperature goes above or below 68 degrees Fahrenheit (20 C), it becomes more important in plant productivity and persistence.

Since temperatures fluctuate all the time, regularly or irregularly, plants must be buffered or be flexible enough to adjust to short-term temperature changes. They also must have adaptive systems that synchronize growth and reproduction with daily and seasonal temperature cycles. Plants generally accommodate temperature extremes by becoming dormant in midsummer or winter, and develop temporary cellular resistance (hardiness) to heat or cold. Many use temperature changes to begin germination and reproduction.

In the Mediterranean region, cool winter and early spring are the most favorable growing seasons; midsummer drought limits plant growth. Plants in this region are adapted to growing actively at moderately low temperatures, but tend to become dormant at high temperatures, especially if water becomes limiting. Since many of the grasses that we use in pastures come from that region, our pasture productivity follows their growth curves. A lot of forage is produced during cool moist weather, but less during hot dry weather.

Grazing management must take into account this variation in plant growth rate during the season. The main requirement of Voisin management intensive grazing is to vary recovery periods according to plant growth rate, to allow plants enough time to regrow and recover from the effects of grazing. Because of this variation, the only way you can manage pasture well is to walk (not drive or ride) over all your pasture area at least once a week and observe how the plants are doing (see chapter 8, Feed Planning).

PLANT ROOTS
Probably because they are out of sight, hardly anyone considers the root systems of plants when managing pastures. But, just as plant tops are affected by selective grazing of variable intensity, roots are also influenced. Since water and nutrients are taken up by roots and nitrogen is fixed in the roots, any beneficial or adverse effects on roots are important to pasture productivity.

Root temperature is always close to soil temperature. If root temperature rises above or falls below the optimum for a plant, growth of that plant slows or stops. The plant immediately is at a disadvantage relative to its neighbors, if they have different genetic traits that enable them to better withstand or use higher or lower root temperatures.

Soil moisture not only affects forage yield, but also how carbohydrates produced in photosynthesis are used. For example, as soil dries, carbohydrates move to roots to support increased root growth, as plants search for water.

The rhizomes and stolons of some pasture plants give

them a competitive advantage, especially in resisting treading by grazing livestock. Plants with deep taproots, (e.g. alfalfa and some weeds) can compete successfully against shallow-rooted plants (e.g. white clover and grasses), especially in drier soil conditions.

Weeds frequently flourish when growth of desirable plants is depressed by adverse conditions or poor grazing management. When desirable plants are grazed in preference to weeds, root systems of the desirable plants become restricted, and weeds gain competitive advantage.

It's interesting to note that about half of pasture plant growth is unavailable to livestock, because it is in the roots. The proportion of plant tops to roots increases if soil nutrients are present in adequate amounts, because the plants don't need to develop such extensive root systems to find and take up the nutrients.

Overgrazing

Andre Voisin showed that overgrazing isn't related to the number of animals grazing, but to the *time* that plants are exposed to them (see chapter 5). Allan Savory built upon Voisin's original observation about time to define overgrazing as any grazing of leaves that regrew from root energy, rather than solar energy. Too frequent grazing to low plant residual without adequate recovery periods, results in reduced pasture productivity. This is because the amount of roots that plants can maintain decreases when they are overgrazed. Overgrazing generally occurs under continuous grazing or inflexible rotational grazing of an insufficient number of paddocks.

After every severe grazing that removes most leaf surface, the plant mobilizes energy that had been stored as carbohydrates in the roots. As carbohydrates are removed from roots to support leaf regrowth, the roots die, separate from the plant, and eventually decompose. Carbohydrates continue to flow from roots to tops until enough leaf surface develops to intercept sufficient solar energy to support further leaf growth, and the regrowth and reestablishment of roots that were lost. If the plant is grazed again before leaves have fully regrown and nearly all roots have reestablished, the plant is overgrazed.

Recovery periods (time required for plant tops and roots to fully regrow after grazing) needed to avoid overgrazing vary according to:
• Growing conditions, as affected by temperature, soil moisture, and nutrient availability.
• Amount of leaf surface remaining after grazing.
• Amount of carbohydrates available from storage.

SOIL ORGANISMS
Botanical composition and productivity of a pasture sward reflect its soil. Quality of a soil depends on the life within it. So in trying to improve pasture productivity, we should concentrate on ways that favor development of soil life.

Microorganisms
Pasture soils contain some of the highest root concentrations of all crops. And there are 20 to 50 times more bacteria and fungi in soil near plant roots (rhizosphere) than in soil away from roots. So the microbiology and chemistry of pasture soils actually concern the rhizosphere, rather than soil alone. In the rhizosphere, chemical soil properties and microorganisms are affected by living plant roots, and vice versa.

Chemistry of the rhizosphere differs from soil away from plant roots because of movement of ions to roots, uptake of nutrient ions by roots, and release of balancing ions and soluble organic materials from roots. These materials include organic acids, sugars, and amino acids. Microorganisms use organic materials as energy sources. Consequently, large populations of microorganisms develop whenever organic materials are available.

Microorganisms in the rhizosphere influence plants in

many, usually helpful ways. Uptake of nutrients (e.g. phosphorus, potassium, molybdenum, manganese) is improved by microorganisms. This probably occurs because microorganisms produce water soluble compounds that help release some nutrients from soil minerals. Plant development proceeds faster when growth factors produced by microorganisms exist in soil.

Some microorganisms in pasture soils inhibit nitrification (oxidation reaction that forms nitrate from ammonium). This means that pasture plants must depend more on ammonium for their nitrogen needs than on nitrate. Consequently, losses of nitrogen through leaching (loss of nitrate) or denitrification (reduction reaction that uses nitric acid from nitrate to form nitrogen gas, which is lost) are minimal under pastures.

Legume-Rhizobium Symbiosis

One of the most important relationships between plants and microorganisms in pastures (and elsewhere), is the symbiosis between legumes and rhizobia bacteria. A symbiosis is a biological relationship between two or more organisms that benefits all of the organisms involved. In this case, the legume-rhizobia symbiosis benefits themselves plus all other organisms. Life as we know it would not be possible without this symbiosis that makes atmospheric nitrogen available to living organisms. Many things affect the legume-rhizobia symbiosis, but let's just look at those that we can influence to help it work better.

In this symbiosis the legume plant determines whether or not it will be nodulated by rhizobia. The plant releases certain organic materials from its roots which develop a rhizosphere that's favorable to rhizobia. Once the rhizobial population builds up, some bacteria infect the plant, and the plant forms a nodule around each bacterium. Within nodules the bacteria multiply and change to a form that is no longer free-living (bacteroid), but must depend on the plant for nutrients to support it.

In return for nutrients and protection, bacteroids fix nitrogen, which becomes available to the plant.

For this symbiosis to function well, all required nutrients must be available in adequate amounts. Besides the nutrients needed by plants, cobalt must also be present because the bacteria require it. The nutrients specifically involved in nodule formation and nitrogen fixation are molybdenum, sulfur, copper, boron, iron, and cobalt. Any nutrient deficiency for the plants reduces their growth and vigor, and results in less nitrogen fixation.

Soil pH must be at the level required by rhizobia for nodulation and nitrogen fixation to occur. Rhizobia differ in their soil pH requirements. For example, rhizobia that infect alfalfa perform best at pH 6.5; those that infect red and white clover tolerate lower soil pH levels, but do best at pH 5.5 to 6.0.

Sunlight affects the symbiosis in at least two ways:
- The main effect involves photosynthesis and the subsequent transfer of photosynthesis products to root nodules. There is a direct relationship between sunlight intensity, nodulation, and nitrogen fixation. Low-light conditions reduce the amount of carbohydrates available to nodules, and nitrogen fixation slows or stops.
- Another effect of light on nitrogen fixation concerns nodule development. Red light is required for nodules to form, but far-red light inhibits nodule development. In pasture swards when legumes are shaded by tall-growing grasses, red light is filtered out by the grasses. If sunlight that reaches legumes contains more far-red than red light, nodule development may stop. Maintaining a favorable red to far-red light ratio for legume growth and nitrogen fixation is one of the reasons for keeping the sward short (less than 6 to 8 inches tall) through grazing management.

Some legumes tolerate shading better than others. If birdsfoot trefoil is only slightly shaded, for example, its nodules drop off. Red clover and alfalfa tolerate shade better than trefoil. White clover tolerates shade better than trefoil, but not as well as alfalfa and red clover.

Adequate soil moisture is needed for nodule formation and nitrogen fixation. If conditions are too dry, fixation slows; if the stress is too severe, nodules drop off.

Too much water has similar effects. If soil is too wet, gas exchange can't occur between nodules and soil air. Low oxygen exchange results in reduced nitrogen fixation and eventual loss of nodules.

Defoliation can greatly affect rate and amount of nitrogen fixation. Legumes may shed all or most of their nodules every time they are grazed or cut. This happens because when leaves are removed, photosynthesis is drastically interrupted, and carbohydrates no longer move to nodules. On the contrary, carbohydrates move upward from storage to regrow leaves. So every time legumes are grazed, their nodules may fall off and decompose. Under favorable conditions, rhizobia reinfect the plants, nodules reform, and nitrogen fixation begins again.

Earthworms

All farmers should actively provide conditions that help maintain or increase earthworm populations for the health of soils in general. Pasture production can be up to 25% greater on soils containing earthworms, compared to soils without them.

Earthworms benefit pasture productivity by:

• Aerating and loosening soil, which enables legumes to fix more nitrogen and encourages plants to root deeper. It also improves water penetration into the soil, retention during drought, and drainage during wet periods.

• Incorporating dead pasture thatch (large amounts of thatch on soil surface indicates low earthworm numbers) into the soil where it is decomposed, thereby recycling nutrients, increasing soil organic matter, and reducing facial eczema spores.

- Breaking down manure quickly, thereby recycling nutrients and reducing fly reproduction sites and internal parasite larvae levels.
- Eating nematodes that can damage clover roots.

Pasture soils contain three to four times more earthworms (about 1,200,000/acre) than tilled soils (400,000/acre). The weight of earthworms in a pasture soil may be more than twice as much per acre as the weight of livestock carried per acre on the surface! In a well developed Vermont pasture we found 1,109,000 and 1,776,000 earthworms/acre, with total liveweights of 2,318 and 3,454 lb/acre, under management intensive grazing by cattle and sheep, respectively. These earthworm weights correspond to a stocking rate of more than 2 to 3 animal units per acre! (1 animal unit = 1000 lb liveweight)

These numbers are directly reflected in the amounts of soil moved. Every day active earthworms take in an amount of soil equal to their body weights. This activity makes a very important contribution to aeration and movement of soils. For example, the total annual excrement of earthworms is about 20 to 30 tons/acre in old permanent pastures, compared to only 9 tons/acre on tilled land. This means that earthworms move at least 20 to 30 tons of soil per acre in permanent pastures each year! As pastures get older and grazing management favors plant growth, earthworm numbers and amount of soil moved increase.

Earthworms not only move and aerate soil, but also make some elements more available for plant growth. When earthworm excrement (casts) is compared to the top 6 inches of soil, the excrement contains five times more nitrate nitrogen, twice as much calcium, almost three times more magnesium, seven times more phosphorus, and 11 times more potassium than the soil. The casts also have a higher pH than surrounding soil.

An adequate level of nutrients, especially nitrogen and calcium in a pasture favors the development of a larger earthworm population. This is probably due to more shading (which keeps the earthworms cool and moist) as a result of better plant growth with available nutrients. Also earthworms have more food at their disposal in the form of fallen leaves and dead roots from the increased plant growth. But don't use urea to apply nitrogen to pasture soil; applying as little as 40 lb urea/acre can cut earthworm numbers in half. Although urea initially costs less than other forms of nitrogen fertilizer, it can lose 40% of its nitrogen by volatilization, resulting in higher actual cost, especially if costs of a decimated earthworm population are considered. Ammonium nitrate is a better source of nitrogen that doesn't volatilize or damage earthworms as much as urea.

Returning animal excrement to the soil increases the number and individual weights of earthworms present in a soil. This is extremely important in a pasture with regard to the breakdown and decomposition of manure. When a pasture environment is in good condition, it contains large numbers of earthworms and other organisms that rapidly break down and decompose the organic matter of manure, releasing the nutrients it contained into the cycle again.

Legumes such as white clover seem to have an especially beneficial relationship with earthworms. Earthworms feed on dead legume residue, and legumes gain from the improved soil fertility due to earthworm excrement.

Soil acidity below pH 5.6 generally is unfavorable to earthworms. Some earthworm families are sensitive to high hydrogen ion concentration of acid soils, and others are sensitive to low calcium availability of acid soils.

Applying herbicides (especially 2,4-D), insecticides (especially those applied against cutworms and corn ear worms), and fungicides can drastically reduce earthworm populations. Earthworm numbers drop rapidly under pesticide application, rotary hoeing, all forms of prolonged cultivation and row-cropping where no animal manure is applied, and any practice that leaves the soil surface bare of plant cover, especially in fall and winter. Plowing the soil kills or weakens earthworms, and the flocks of gulls and other birds that follow the plow eat exposed worms, further decreasing the earthworm population.

Earthworms must have plant cover for food during the entire year, and for protection during the fall. Earthworms remain active all winter, and must have a period of acclimation in the fall so they can survive freezing temperatures. Plant cover on the soil surface insulates them until they develop cold resistance.

If little or no decaying plant or animal residue exists on the soil surface, there is no reason for surface-feeding earthworms to burrow their way to the surface. Surface feeders decline in favor of earthworms better adapted to eating buried, decaying residue. So the channels (macropores) from surface to deep in the soil that surface feeders would have formed in the process, don't get formed enough to allow water to penetrate, and the water cycle deteriorates. With each rainfall or snowmelt, successively more runoff occurs, increasing topsoil erosion onto lowland and into streams, and flooding of farmland and cities along rivers.

Common spring floods and the massive flooding that occurs along the Mississippi River may be a symptom of the poor condition of American farmland. These disasters may be linked to practices that greatly simplified plant and animal populations, mismanaged grazing, or totally removed grazing animals, resulting in decreased numbers and kinds of soil organisms, including earthworms.

Nematodes

Despite their poor image, nematodes are beneficial and important, due to their vast numbers in soils. (We found 2,810 to 15,208 nematodes per 100 grams of soil under well developed pasture in Vermont!) Activities of dominant nematodes result in rapid decay and incorporation of organic matter within the soil, and nutrient cycling. Some nematodes feed on bacteria, fungi, and soil protozoa, thereby assisting in the natural balance of other soil life forms. Destructive or pathogenic nematodes are absent or present in very low numbers of less than 1 percent of the nematode population in nontilled soils. When they are present in soils in good condition, pathogenic forms are kept in check by predatory nematodes.

PESTS

The presence of a large number of pests of any kind in a pasture is usually a symptom of poor management that interferes with the natural balanced order of the ecosystem. Problems arise when we do something to part of an environment, without having the slightest idea about the possible effects that our interference might have on the whole ecosystem.

Weeds

There really are no weeds in a properly managed pasture, except for noxious plants such as thistles and poisonous plants that simply can't be grazed without harming the livestock. But we interfere with the pasture environment when we under- or overstock an area and continuously graze it with our livestock, that aren't free to move on after they have grazed it once, as they would do under natural conditions; this brings on all kinds of weeds.

Forbs

Animals graze first and repeatedly plants that they like best, such as clovers, lush grasses, and dandelions. They leave forbs such as smartweed and pigweed because the plants have a smell or taste that animals dislike, or they are covered with short fuzzes that tickle their mouths. Despite being a little less palatable, these so-called weeds have high feeding values when immature. If they aren't eaten along with clovers and grasses, the "weeds" gain competitive advantage. They mature, multiply, and spread throughout the pasture, progressively decreasing the amount of more desirable plants in the sward.

In an overstocked continuously grazed pasture, animals graze all plants repeatedly, except noxious weeds, every time the plants grow tall enough to be grasped by the animals. By early July such pastures appear to be producing nothing to sustain the animals. Under such grazing pressure many desirable plants assume an extremely low-growing form that allows them to escape being grazed. White clover, for example, will produce flowers 1 inch or less from the soil surface, depending on whether they're grazed continuously by cattle or sheep. Less desirable or harmful plants take over the pasture, since all of their competition is removed by the grazing.

Noxious Weeds

Usually pastures are terrific messes by the time people begin using management intensive grazing. Three kinds of noxious weeds may be present in your pasture (see Appendix list of noxious weeds and their control):
• Weeds that can cause mechanical injury to livestock are fairly common. They have hard, sharp parts (e.g. awns, spines, thorns) that can pierce or cut. Wounds caused by these plants around mouths and eyes may be painful and also may enable bacterial infections to start, resulting in swelling and pus. Affected animals may eat less or starve because of blindness or eyes swollen so shut that they can't find forage. Sharp parts of eaten plants may cause serious internal wounds or obstructions in the digestive tract.
• Strongly flavored weeds can cause off-flavor in milk of animals that graze them; affected dairy products have less or no value.
• Poisonous plants must be treated carefully. Losses of livestock from eating poisonous plants are relatively

low in the eastern USA, but can be quite serious in western areas, where animals graze more native vegetation. Don't force animals to graze down any area that contains plants that are poisonous to livestock. Normally animals know better than to eat such plants, but if you starve them into eating poisonous plants, they will do it. Remove or exclude poisonous plants from the pasture before you have your animals graze it closely.

General Weed Control
The presence of certain weeds indicates soil fertility, moisture, or pH problems. For example, acid soil and need of liming is indicated by presence of sorrels, dock, lady's thumb, horsetail, hawkweed, and knapweed. Crust formation and/or soil hardpan is indicated by field mustard, horse nettle, penny cress, morning glory, knotweed, and quackgrass. Heather and knotweed indicate soil acidity and low phosphorus and potassium. Changing these soil conditions by liming, fertilizing, draining, or irrigating to favor desirable plants can eliminate weeds.

Almost any plant that protects itself from being grazed by having thorns (e.g. thistle), fuzzes (e.g. pigweed), bad taste (e.g. milkweed, burdock), bad smell (e.g. skunk cabbage), or poison (e.g. bracken fern) usually can be eliminated simply by cutting it off. You can see that this is true by looking in hay fields and lawns. Usually these kinds of noxious plants don't exist where land is periodically mowed.

The most dependable, accurate, easy-to-use herbicides for controlling noxious weeds in pasture are mowers (rotary, flail, or sickle), scythes, and machetes. With these there are no residues in forage, and instead of adverse side effects, mowing a pasture improves general forage production, quality, and use.

To control noxious weeds, cut them off as close to the soil surface as possible when they form buds, before flowering and seed set, so they don't reproduce. At this stage much of their food reserves have been used to form buds and flowering stems, and removing their leaves weakens or kills them. They may have to be cut off repeatedly, because some of them can regrow from remaining root reserves. Usually two or three cuttings will eliminate even the most persistent species. Intensive management grazing favors desirable plants, which will become stronger and eventually compete successfully against reestablishment of noxious weeds.

Careful use of chemical herbicides (e.g. glyphosate) to individually treat noxious weeds may eliminate them quicker than by cutting. One way of treating weeds individually is with a wick or mop applicator. Herbicide is placed in the tube handle and weeds are touched with the wick or mop (Available from Oldfields Seed, 800-448-5145 or Taylor Farm, 912-849-3855). A weed wiper is available for applying chemical herbicide to large areas of weeds without spraying (Doomsday to Weeds, 813-949-5579).

Spraying large pasture areas with chemical herbicides is not advisable because they affect all plants, not just the ones you want to get rid of. Chemical herbicides also affect soil organisms in complex ways that may or may not be to your benefit. Besides, herbicides and other pesticides are hazardous to the health of all organisms, including you. First try cultural practices, such as changing your grazing management and/or mowing, to resolve weed problems. If you must use chemical herbicides, be extremely careful!

Insects
Insects of all kinds (locusts, grasshoppers, crickets, leafhoppers, aphids, weevils, grubs, termites) live in pastures and damage them to some extent. The damage results mainly from them eating leaves, boring into stems, sucking plant juices, introducing fungi and viruses, and eating roots.

In balanced environments, insect damage isn't severe or important. Birds, diseases, parasites, other predators, and adverse weather all help control insect populations.

Insect problems in pastures can be brought on by things we do that change conditions of the pasture area. For example, overgrazed (e.g. continuously grazed) pastures provide ideal situations for breeding of grasshoppers and locusts. When woodlands are reduced and ponds and marshes drained, bird populations are affected, and conditions may become favorable for development of larger insect populations. It has been estimated that 15 locusts or grasshoppers per square yard of a 40-acre pasture eat the equivalent of 1 ton of hay each day! So it certainly is worthwhile to do things (e.g. provide birdhouses) that encourage large numbers of insect-eating birds to live in or near pastures. Birds also help break up cattle dung and speed its decomposition, by digging in the dung for seeds and insect larvae. (Raising poultry on pasture, with or following livestock, provides similar benefits.)

Flies can cause serious problems to grazing livestock. Like other insect pests, however, flies decrease in number in properly managed pastures. This secondary benefit of good pasture management may result from more uniform distribution of manure, and its rapid breakdown and decomposition. Large bird populations also may help decrease fly problems by eating flies and their larvae.

Diseases
About 45 diseases can affect pasture plants in the United States. Grazing management that results in a mixed population of vigorously growing plants probably is the best control of these diseases, which tend to infect weak plants. Grazing pastures down closely in fall or winter reduces overwintering of disease organisms on leaves; this helps decrease reinfection of plants in spring.

Rabbits, Hares, And Rodents
Rabbits, hares, and various rodents (prairie dogs, woodchucks, gophers, mice, rats, ground squirrels) often

seriously damage pastures and rangelands. Overgrazing creates conditions that favor development of high populations of these animals, and thereby contributes to increasing the damage done by them. Destruction of useful wildlife (wolves, coyotes, wild cats, owls, eagles, hawks, and many snakes) also allows rabbit, hare, and rodent populations to increase excessively. In California, for example, it was estimated that ground squirrels were eating as much forage each year as 160,000 cattle! Encouraging abundant diversified wildlife populations, and properly managing pastures are the best ways to minimize damage from these animals.

SHADE FOR LIVESTOCK
Everyone is concerned about livestock when the weather is hot and there's no shade available for them. Simply providing shade, however, may not avoid decline in production during hot weather, because other things may be involved:
- Inadequate availability of high-quality water. During hot weather animals eat less forage and drink more water. If they can't drink enough water, forage intake decreases. Poor water quality can affect animals more than heat; high temperatures worsen a problem of poor water quality. A good test of water quality is: would you drink it? If water quality is poor (warm, hot, dirty, algae in tanks) and/or animals have to walk more than 600 feet to drink, production will decrease.
- Low sodium and/or magnesium in total ration.
- Grass with high endophyte content.
- Inadequate selenium in diet in total ration.
- Flies and midges that pester livestock.
- Inadequate forage allowance.
- Reduced forage intake of hot and wilted plants.
- Decreased forage intake if clover content of sward is low.
- Lower forage palatability and intake due to soil (dust) on plants or pulled plants, or because of decomposing plant

residue from too much pre- and postgrazing mass. Smell the forage; if it's moldy don't force animals to graze so closely. Change grazing management in the spring to avoid buildup of moldy material by midsummer.

Adverse effects of heat stress can be reduced greatly by:

- Making certain that animals are well fed; increase forage allowance as needed. Allocate more forage after the evening milking than after the morning milking, because during hot weather animals will eat more in the evening.
- Providing clean, cool, high-quality drinking water in paddocks or nearby. Clean water tanks as needed to prevent buildup of algae.
- Providing adequate amounts of minerals, especially salt. Sodium in salt helps animals (and people) control body temperature, so it's more important than shade. Always have salt and minerals available free-choice in loose form, not solid blocks so livestock can get enough to meet their needs. Another way to provide minerals is by metering them through the water supply.
- Moving dairy animals and milk during cooler times of the day.
- Moving dry stock and growing animals in the evening.

Heat stress is worse for animals when they are concentrated. One place where this occurs daily is when dairy cows stand in a hot barnyard waiting to be milked. This situation should be improved by installing overhead lattice screens for partial shade and air movement, and fine-mist sprinklers to cool the cows. It's best to start the sprinklers to cool the yard before the cows get there.

Providing shade for grazing animals isn't a simple matter. If a pasture has too many trees, forage production can be improved by thinning the trees to cover about 30 percent of the area, with as uniform a distribution as possible (no closer than 30 feet apart in all directions). This amount of trees provides adequate shade, doesn't interfere with solar energy interception by pasture plants, and doesn't result in as much concentration of nutrients from manure as usually happens when only a few trees are present in a pasture.

Pastures typically have two or three trees or groups of trees at various locations. Animals tend to graze away from the trees, but camp around them. Although this transfers nutrients to the immediate area around the trees from the rest of the pasture, and in the long term decreases pasture productivity unless you apply manure or fertilizer, it's a small price to pay for having valuable trees on your land. Earth needs all the trees that can grow.

Establishing trees in a pasture to provide uniform shade is neither easy nor inexpensive. Besides the cost of buying and planting trees, each sapling must be protected from being eaten by grazing livestock. I've tried various ways of establishing trees on pasture, including spraying saplings with manure tea and using tree shelter metal mesh or plastic tubes (e.g. Tubex tree shelters from Trident Enterprises: 800-533-0213), with and without electric fence. Manure tea seems to make saplings more attractive to livestock! Without electric fence, animals rub on the tree shelter tubes until the tubes are knocked over, and then eat the saplings. One problem I've had with shelter tubes is that bluebirds (!) tend to get trapped and die in them; to avoid this, cover tube tops with 1/4 inch metal mesh.

The only way that I've found that works is to plant saplings in rows, with or without tree shelter tubes. Both sides of each row must be protected with electric fence that allows livestock to graze around the saplings without touching them. The rows should begin at paddock perimeter fences and extend only partly across paddocks, so animals can move around the rows of growing trees. Single-strand fence can be used to protect trees from cattle and horses, but multiple-strand fence is needed to protect them from goats and sheep.

If you decide to plant trees on pasture to provide shade, you should consider choosing tree species that allow you to double crop trees and pasture (see Smith's *Tree Crops* in References). For example, black walnut trees produce extremely valuable lumber; honey locust trees grow quickly, fix nitrogen, and provide nutritious pods for animal feed in autumn.

Due to lack of shade, some dairy farmers let cows out to graze in the morning, and keep them in the barn during the heat of the afternoon. This drains nutrients from pasture even worse than a couple of shade trees in the pasture. It also requires powered ventilation and some feeding of the animals in the barn (which substitutes for pasture forage they otherwise would have eaten), as well as cleaning the barn and spreading manure -- all take time, labor, and energy.

There's another way to work around heat and shade problems during hot weather. Try grazing paddocks that have shade during the daytime, and at night graze paddocks that don't have shade.

Probably the best way is to only have animals that are well suited to your local environment and tolerate heat. For example, brown-colored dairy breeds (e.g. Jersey, Ayrshire, Brown Swiss) continue grazing long after Holsteins have given up during hot weather.

SHELTERBELTS OR WINDBREAKS
Planting shelterbelts or windbreaks perpendicular to the prevailing wind at least every 600 feet along paddock division fences can provide many benefits:
• Prevent wind erosion of topsoil.
• Shade for livestock.
• Shelter outwintered livestock, saving money in reduced feed needs because sheltered livestock use less energy to maintain body temperature.
• Greater pasture production from more effective moisture (lower evapotranspiration) and less stress on plants due to reduced wind velocity.
• Increased available soil moisture from dew formation

on tree leaves in summer, and snow settling out in shelterbelts in winter.

- Longer growing season due to higher temperature of sheltered land.
- Enhance habitat of insect-eating birds.
- Reduce losses of fruit and grain crops caused by wind.
- Reduce home heating costs by lowering wind chill.
- Improve living and working conditions around farm.

Height, porosity, length, and uniformity influence the effectiveness of shelterbelts. Height determines the size of area that wind speed is reduced on leeward and windward sides. Porosity determines how the wind behaves on the leeward side. Dense, impermeable windbreaks provide the most protection 2 to 5 times the tree height. Protection drops quickly because negative pressure forms downwind; this pressure difference also dries the soil. A great deal of turbulence also may develop on both sides of impermeable windbreaks. Shelterbelts with medium porosity (40-60%) result in more even wind flow than dense windbreaks, and although initial protection isn't as great, it continues out 25 to 30 times the height.

Minimum length of a shelterbelt should be 20 to 25 times its mature height. This relationship is needed to maintain protection when wind changes from right angles to the shelterbelt or when wind speed is high and wind could move around the ends of a short shelterbelt.

Good management is needed during establishment of shrubs and trees to prevent formation of gaps in shelterbelts. Any gaps should be replanted as soon as possible; otherwise wind funneling through the gaps later will prevent establishing new trees.

Be careful to orient shelterbelts to avoid shading and forming frost pockets. If possible, shelterbelts should run North-South so sun shines on both sides. If you need East-West shelterbelts, use only deciduous trees in them to allow sunlight through as much as possible.

Even a single row of conifers can provide a warmer

environment by blocking the wind and reducing the chill factor. Multiple rows are needed, though, in snowy areas to catch snow in shelterbelts, not on leeward sides.

Because of their fast growth rate, hybrid poplars in shelterbelts quickly provide some protection from wind during winter, and slow the wind and provide shade during the growing season. A row of berry bushes in shelterbelts makes them valuable wildlife habitat areas that will provide immeasurable benefits to you. One thing that you'll notice very quickly are the greater numbers of birds and bird species around your farm because of the increased cover and food supplies.

Some tree species can be included in shelterbelts for increasing income and pasture productivity. For example:

• Black walnut produces valuable lumber and nuts.
• Black locust grows rapidly, fixes nitrogen, and cycles nutrients from deep in the soil. It is excellent for honey production and produces dense, rot-resistant wood for posts and firewood; trees regrow after cutting.
• Honey locust trees produce pods that are highly palatable and nourishing to livestock.

These trees are especially useful in East-West shelterbelts to allow sunlight to reach pasture plants on both sides. The trees leaf out late in spring and loose leaves early in fall; they also have open canopies that allow sunlight through while in leaf.

Many trees and bushes are available for shelterbelts:

• <u>Conifers</u>:

Tall (50-75 feet) : red cedar, ponderosa pine, scotch pine, Norway spruce, Douglas fir.

Medium (20-50 feet) : Colorado blue spruce, Black Hills spruce, Austrian pine, northern white cedar, eastern red cedar, Rocky Mountain juniper.

• <u>Deciduous trees</u>:

Tall (50-75 feet) : hybrid poplar, Lombardy poplar, black walnut, hackberry, green ash, Chinese elm, Siberian elm, Sioux cottonwood, silver maple, black locust.

Medium (20-30 feet) : honey locust, Russian olive, willow, caragana.
• Shrubs *(8-15 feet)* : black elderberry, cardinal autumn olive, common lilac, highbush cranberry, honeysuckle, plum, Siberian peashrub.

The main or first windward shelterbelt should be three or four rows wide of shrubs (1 row), deciduous trees 1-2 rows), evergreen trees 1-2 rows). The row of shrubs should face the prevailing wind, with taller trees in the middle, to form a wedge for lifting the wind up and over the area. Single-row windbreaks of alternating evergreen and shrubs can be used for supplemental windbreaks every 500 to 600 feet. It's best to fence livestock out of shelterbelts to protect the saplings, and create wildlife habitats.

Give your trees room to grow, even if it means that you'll have to use fewer rows. Preferably, rows should be planted 16 feet apart. Spacing within rows of multiple-row windbreaks should be: tall deciduous and evergreen trees, 12 feet apart; medium deciduous and evergreen trees, 9 feet; and bushes, 4 to 6 feet. Spacing within single-row windbreaks should be: tall trees, 8 feet apart; medium trees, 6 feet; and bushes, 2 to 4 feet.

Help your trees to grow by mulching each tree to reduce grass competition. I use two catalogs or two newspapers to mulch each tree. Just tear a catalog or complete newspaper edition about one-fourth of the way through across a long side, slip the torn opening around the sapling. Then lay another catalog or newspaper edition over the torn ends of first one that extend on the other side of the sapling. That's all there's to it; the mulch will last for at least a couple of years.

Some sources of trees and shrubs for windbreaks are:
• Gurney's, 110 Capital St., Yankton, SD 57079, phone: 605-665-1930; specializes in shrubs and trees for windbreaks.
• J.W. Jung, Randolph, WI 53957, phone: 800-247-5864; has an excellent hybrid poplar and shrubs for birds.

REFERENCES

Bergersen, F.J. 1982. *Root Nodules of Legumes: Structures and Functions.* Res Studies Press, London.

Davidson, R.L. 1978. Root systems--the forgotten component of pastures. p. 86-94. In J.R. Wilson (ed) *Plant Relations in Pastures.* CSIRO, Melbourne, Australia.

Feldrake, C.M. 1994. Agroforestry: can trees improve profit from pasture? Stockman Grass Farmer, 51(6):12.

Hardy, R.W.F. 1977. *A Treatise on Dinitrogen Fixation. II. Agronomy and Ecology.* John Wiley and Sons, Inc., NY.

Harper, J.L. 1978. Plant relations in pastures. p. 3-16. In J.R. Wilson (ed) *Plant Relations in Pastures.* CSIRO, Melbourne, Australia.

Harris, W. 1978. Defoliation as a determinant of the growth, persistence and composition of pasture. p. 67-85. In J.R. Wilson (ed) *Plant Relations in Pastures.* CSIRO, Melbourne, Australia.

Jones, V. 1992. Shade, heat & cold. *Stockman Grass Farmer.* 49(2):28.

Jones, V. 1993. Everything (and then some) you ever wanted to know about earthworms. *Stockman Grass Farmer.* 50(6):10-11.

Jones, V. 1995. Reducing the summer slump. Stockman Grass Farmer, 52(7):25.

Kemper, W.D., and T.J. Trout. 1987. Worms and water. *J. Soil & Water Conservation.* Nov.-Dec. p. 401-404.

Ludlow, M.M. 1978. Light relations of pasture plants. p. 35-49. In J.R. Wilson (ed) *Plant Relations in Pastures.* CSIRO, Melbourne.

McWilliam, J.R. 1978. Response of pasture plants to temperature. In J.R. Wilson (ed) *Plant Relations in Pastures.* CSIRO, Melbourne.

Mollison, B., and R.M. Slay. 1991. *Introduction to Permaculture.* Tagari Publications, Tyalgum, Australia. 198 p.

Mortimer, J. & B. 1997. Well-planned tree plantings can cut shelter capital costs. Stockman Grass Farmer, 54(1):1.

Nation, A. 1994. Some tips on barnless wintering. Stockman Grass Farmer, 51(11):10.

Nation, A. 1996. Vermont graziers get pasture management tips from Ireland. Stockman Grass Farmer. 53:(10):6.

Pacific Northwest Cooperative Extension. 1975. *Trees Against the Wind*. PNW Bulletin No. 5. 47 p.

Pfeiffer, E.E. 1990. *Weeds and What They Tell*. Bio-Dynamic Farming and Gardening Association, PO Box 550, Kimberton, PA 19442.

Rhodes, I. and W.R. Stern. 1978. Competition for light. p. 175-189. In J.R. Wilson (ed) *Plant Relations in Pastures*. CSIRO, Melbourne.

Ronneberg, E. 1994. How about some trees? Stockman Grass Farmer, 51(6):17.

Rovira, A.D. 1978. Microbiology of pasture soils and some effects of microorganisms on pasture plants. p. 95-110. In J.R. Wilson (ed) *Plant Relations in Pastures*. CSIRO, Melbourne, Australia.

Savory, A. 1988. *Holistic Resource Management*. Island Press, Washington, DC. 565 pp.

Savory, A. 1992. Time & overgrazing. *HRM Newsletter*. 37:5.

Semple, A.T. 1970. *Grassland Improvement*. Leonard Hill, London.

Smith, B., P.S. Leung, and G. Love. 1986. *Intensive Grazing Management*. Graziers Hui, Kamuela, Hawaii. 350 p.

Smith, J. R. 1929. *Tree Crops*. Harper & Row, NY. 408 p.

Turner, N.C. and J.E. Begg. 1978. Responses of pasture plants to water deficits. In J.R. Wilson (ed) *Plant Relations in Pastures*. CSIRO.

Vincent, J.M. 1974. Root-nodule symbiosis with Rhizobium. p. 265-341. In A. Quispel (ed) *Biology of Nitrogen Fixation*. North Holland.

Voisin, A. 1959. *Grass Productivity*. Philosophical Library, NY. 353 p.

Voisin, A. 1960. *Better Grassland Sward*. Crosby Lockwood, London.

3
Pasture Nutrition

The morning glory that blooms for an hour
Differs not at heart from the giant pine,
Which lives for a thousand years.
                                    Anonymous

As far as we know, plants require 22 elements to live and grow. These are: boron, calcium, carbon, chlorine, cobalt, copper, hydrogen, iodine, iron, magnesium, manganese, molybdenum, nickel, nitrogen, oxygen, phosphorus, potassium, selenium, silicon, sodium, sulfur, and zinc. Notice that I've listed them in alphabetical order. Any other order would imply that some are more important than others, when actually all are equally important to plants, but needed in different amounts. Plants take at least these elements and, with the driving force of solar energy, integrate them into living life forms. If any one of them is absent, life is absent. All life therefore derives from the sun, and plants' ability to use solar energy to organize these elements.

All of the elements except carbon, hydrogen, nitrogen, and oxygen ultimately derive from the parent material that developed into soil. Carbon, hydrogen, and oxygen come from the atmosphere and water. Nitrogen comes from the atmosphere through biological nitrogen fixation. The presence of these mineral nutrients and plants' success in obtaining them mainly determine natural distributions of plants, and their ability to grow and survive in different environments.

Soil Fertility
One of the first steps in improving your pasture's productivity is to estimate its soil fertility status. No matter what grazing management is used, plants can't grow if they lack nutrients.

Soil and Plant Tissue Tests

The main way of estimating soil fertility level is to sample and test soils and/or plant tissues, and follow the recommendations that result from the tests. Soil and tissue tests aren't perfect, especially soil tests for pasture, since they were developed using tilled cropland soils. Tissue tests would be preferable in estimating soil nutrient status, because they measure the available elements that growing plants actually extract from a soil. Unfortunately, amounts of elements in plants vary as growing conditions change and as plants mature. Also, almost no research has been done to calibrate plant tissue element content with soil fertilizer and lime needs. Despite problems, soil and plant tissue tests are the best tools we have for making sure that no major soil fertility or pH problems exist.

Follow the testing lab's instructions for taking and handling soil and plant-tissue samples. Use great care in taking tissue samples because the slightest amount of contamination can make test results worthless. Test results can only be as good as the sample analyzed, so your most important part in testing is to take samples that represent the pasture soils and plants.

Since more than 85 percent of pasture plant roots are concentrated in the top 2 to 3 inches of soil, the top layer is most important in plant nutrition and in affecting sward composition and forage yield. The top layer is especially important in permanent pastures, because the soil isn't mixed with fertilizer and lime throughout a 6- to 8-inch depth as in tilled soils. As a result, fertilizer and lime applications to permanent pasture mainly affect this top soil layer. So we should sample only the top 3 inches of permanent pasture soil for testing.

The best way to take soil samples is with a soil probe (auger or tube). You can either buy a soil probe at a farm-supply store or through a farm-supply catalog (e.g. Nasco), or sometimes borrow one from your county Extension office. When using a tube soil probe, just push the point into the soil down to 3 inches, and pull it out straight up, without twisting the probe. If you twist a tube probe, it will break off. Augers are more expensive than tube probes, and aren't needed unless you have extremely stony or hard soils. When pulling a probe (especially augers) out of the soil, lift it out with your knees, never with your back.

If you don't have a soil probe, you can use a shovel. Just push the shovel in about 3 inches and pry the slit apart enough to get your hand in. Then scrape a handful of soil from the middle of the opening, starting from the bottom and scraping upward with your fingers.

If your pasture has obvious soil differences, such as hilltops, hillsides, and valleys, sample the different soils separately and have them analyzed as separate samples. By this I don't mean that each hilltop should be a separate sample, but that all or several hilltops should sampled and combined together as one hilltop sample. The reason for this is that each soil type has different fertility characteristics, especially in pastures, where some areas may have received more dung and urine than others, because of the way that animals graze pastureland.

When sampling an area, move across it in a zigzag pattern, taking samples at the point of each zig and zag and midway between the two. Be especially careful to avoid sampling in or near obvious dung and urine patches. Collect 15 to 20 samples per 5 acres (composite sample), placing them in a clean plastic bucket that never has had any fertilizer in it. Then mix the samples together well, remove about a one-half pound subsample, and place it in a clean plastic bag. Use a permanent marker to label the bag. So that you know where the sample came from, either keep notes or draw a map of the pasture, indicating where the samples were taken. Empty out the bucket and begin taking the next composite sample. Keep soil subsamples in the shade until you have finished sampling, and refrigerated until they can be analyzed.

Follow the testing lab's fertilizer and liming recommendations, or calculate your own rates from the lab's analyses (see *The Farmer's Fertilizer Handbook* in References).

Plant Symptoms
Because of variation in soil and plant tissues, tests may not reveal deficiencies of certain nutrients. Also tests won't indicate anything about soil conditions that adversely affect plants. Besides the tests, you need to closely observe growing pasture plants to make sure they're OK. Plants can show symptoms of nutrient deficiency and/or adverse soil conditions. White clover generally shows easily identifiable, pronounced symptoms before grasses do.

Clover Symptoms
• Black spots under leaves: low phosphorus.
• Brown-colored leaves: low phosphorus.
• Brown spots through leaves: low potassium.
• Hard, dry brown leaves with split and/or cut edges: low boron.
• Poor nodulation: compacted soil, low pH, general low fertility, low molybdenum and/or cobalt.
• Pale, white leaves and stems or parts of them: severe molybdenum deficiency.
• Small leaves: low calcium, phosphorus, potassium, copper, molybdenum, and/or cobalt; or it could be a variety with small leaves.
• Yellow leaves: wet soil, low sulfur.
• Yellow or orange mottling of leaves: severe molybdenum deficiency; symptoms usually appear first in older leaves, then in younger leaves.
• Yellow stripes on leaves: low magnesium or sulfur; or it could be just a seasonal growth habit.

Grass Symptoms

- Hard leaves: excess potassium, low calcium, low sodium, or just a variety with hard leaves.
- Purpling of leaves: low phosphorus; appears more in cold, stressed conditions.
- Shallow roots and pulling out roots: high aluminum, soil hardpan, applying too much soluble phosphorus or nitrogen fertilizer, low calcium, too hard cattle grazing.

Most nutrients remain in a well-managed pasture, cycling continuously through the environment. For example, cattle excrete 80 to 90% of the nutrients they eat. So usually all that's needed in any pasture are adequate levels of nutrients to start with. Keep in mind that any forage, grain, and mineral supplements fed to pastured livestock, in effect fertilize the pasture with nutrients from the supplements that pass through in dung and urine. Generally there's no need for massive annual maintenance applications of any nutrients to well-managed pastures; any fertilizer and lime that's needed is best applied frequently in small amounts.

Now let's look at what happens to nutrients in a pasture environment, to understand how we might influence their availability, distribution, and flow to benefit pasture plants and grazing livestock.

NITROGEN (N)

A grass-legume pasture gets its nitrogen almost entirely from the legumes through biological nitrogen fixation. What does this mean? Well, to begin with, the amount of nitrogen needed each year in a pasture is simply amazing. Take for example a seasonal pasture yield of 5 tons of dry forage per acre. Since this forage contains about 4 percent nitrogen, it means that 400 lb N/acre had to be available to the plants. If we also take into account nitrogen needed for root formation, and nitrogen losses by leaching and denitrification, the total amount of nitrogen available to the pasture environment had to be more than 555 lb/acre during the growing season!

Nitrogen Cycle
Two main pathways exist in which nitrogen transfers from legumes to associated grasses: aboveground and underground.

Most nitrogen transfers aboveground, as nitrogen excreted by grazing animals. The amount of nitrogen removed from pastures by grazing animals is small. For example a wool fleece contains only 1.5 pounds of nitrogen, and a fat lamb contains about 4 pounds of nitrogen. These are tiny amounts compared to the total amount of nitrogen cycling in high-producing pastures.

Besides the small amounts of nitrogen removed from the pasture in livestock products, 70 percent of the rest of the nitrogen ingested in forage is excreted in urine and 30 percent in dung by grazing animals. Nitrogen in urine mainly exists as urea (70%) and amino acids; all are quickly converted to readily available ammonium and nitrate by soil microorganisms.

Livestock grazing a pasture containing only 30 percent white clover, apply nitrogen in urine patches at a rate of about 225 lb/acre. Put another way, 100 cows grazing 1-acre paddocks for 24 hours each, apply 35 to 43 lb of nitrogen, 10 to 23 lb of phosphorus, and 28 to 35 lb of potassium in their urine and dung per acre per day! If you consider other nutrients and organic matter applied by grazing animals, the true value of manure can be appreciated.

In contrast to nitrogen in urine, nitrogen in dung isn't immediately available, because it exists in organic combinations that first must be broken down and mineralized by birds, insects, earthworms, and other soil organisms.

Two other processes contribute to aboveground nitrogen transfer: leaching of nitrogenous compounds from plant shoots, and decay of leaves and petioles that have fallen from plants. In an ungrazed situation such as a hayfield, these are the only ways that nitrogen transfers aboveground to associated grasses. Much less nitrogen transfers in these ways, compared to that transferred in manure.

Underground nitrogen transfer occurs through the leaking of nitrogenous compounds from legume nodules, and the sloughing off and decay of nodules and root tissue. Since white clover roots and nodules contain 1.5 and 6 percent nitrogen, respectively, a lot of nitrogen can be transferred to grasses underground, but not nearly as much as transfers aboveground with grazing animals

Unfortunately, grazing animals don't distribute dung and urine evenly over the pasture, and this is the main reason for losses from the nitrogen cycle. Animals tend to lie down or camp on certain spots within a pasture. The more irregular the pasture terrain, the more the animals congregate in their favorite spots. This effect can be seen especially in hilly country, where livestock graze on slopes and valleys during the day, but spend nights on hill crests. Anyone who has watched animals at all, knows that when they get up after lying down, the first thing they do is poop and pee. If they are always allowed to camp in their favorite areas, nutrients will be transferred from the rest of the pasture to those spots. Over time, excessive nutrient levels accumulate in those spots, while the rest of the pasture becomes deficient in nutrients.

The amount of nitrogen excreted not only varies over different parts of a paddock, but also varies within the patches of dung and urine that are deposited. For example, one urination of a beef cow affects a more or less circular area of 21 to 25 inches in diameter, or 2.4 to 3.4 square feet. Most of the nitrogen in a urine patch is concentrated near the center of the patch. Because of both the uneven distribution of urine throughout a pasture and within urine patches, nitrogen levels in the soil in livestock camps and within urine patches may get as high as 1,500 lb/acre! Even grasses with their large capacity to absorb inorganic nitrogen are unable to take up all the nitrogen present. This leaves some ammonium nitrogen available for conversion through nitrite to nitrate by soil microorganisms. Unfortunately, nitrate nitrogen is easily leached from soil, and, besides being lost to the pasture environment, this nitrate nitrogen can pollute surface and ground water.

Another form of nitrogen loss from urine patches occurs through denitrification, in which nitrogen gas is produced by microorganisms from nitrites and nitrates. Also, about 12 percent of the nitrogen present in a urine patch escapes as ammonia gas, giving urine its characteristic smell. The nitrogen gases return to the atmosphere, possibly to be fixed again in legume nodules and reenter the pasture environment.

Although nothing can be done to even out the distribution of nitrogen within urine patches, management intensive grazing helps to minimize the concentration of manure in livestock camps. Dividing pastures into paddocks and consequently increasing stocking density within paddocks, reduces the concentration of excreta in camp sites. As stocking density increases, less manure is deposited in camp areas, possibly because animals spend less time at any one camp site, as they graze more evenly over a paddock.

Dividing pastures into paddocks doesn't eliminate the problem of nutrient transfer to camp sites, however, because animals soon find preferred camping spots within each paddock. Despite this tendency, the concentration of nutrients is less because they are only being concentrated within the smaller paddock area, and not from the entire pasture.

Try to divide pastures so that spots likely to be preferred for camps are separated from areas where just grazing occurs. For example, try to place hill crests in paddocks separate from slopes or level areas. Also, slopes of different aspect should be separated; that is, shady slope faces should be separate from sunny slope faces, which likely would be camp areas. But if this isn't possible, don't worry about it. Probably the biggest beneficial affect on excrement distribution comes from the increased stocking density per unit area that occurs when pastures are divided into paddocks.

Nitrogen And Grass-Clover Balance

Nitrogen applied in fertilizer, grazing animal excrement, or spread manure to a pasture sward suppresses clover growth, as a result of the increased competition from associated grasses. As discussed in chapter 2, competition may occur for light, water, and nutrients, but usually involves more than one of these. This is because competition for a nutrient or for water has the secondary effect of resulting in different heights of the competing plants, which causes shading of lower growing ones and, therefore, competition for light. Under most pasture conditions, available nitrogen supply limits grass growth. So when nitrogen is applied, grasses use it to produce a large amount of topgrowth that shades associated clover plants, unless grazing management stops it.

Shading restricts root growth more than top growth. Shaded plants then have less capacity to use water and nutrient supplies. This restriction further limits top growth and light interception. The effect of nitrogen application in suppressing clover growth in a pasture, therefore, is due to the nitrogen causing an increase in both light and nutrient competition by associated grasses.

Applying nutrients other than nitrogen that previously limited clovers, usually gives a spectacular burst of clover growth. This is because in the first stages of a grass-legume sward's response to applied nutrients, lack of nitrogen still limits grass growth, and clover competes best for light, water, and nutrients. As soil nitrogen builds up by underground or aboveground transfer of nitrogen from vigorously growing clover, grasses become less limited by nitrogen level and begin to compete more for light, water, and nutrients.

Although legumes have a very high light requirement, it doesn't seem to be for physiological reasons. It's probably due to the plants' growth habit, especially that of white clover. White clover's low-growing habit with a very dense, almost single layered leaf canopy isn't very efficient in intercepting light, compared to the more erect growth of grasses. The suppression of clovers by nitrogen application or buildup can be reduced by well-managed, close grazing to lessen the advantage of grasses in competing for light.

Competition between grasses and legumes occurs for other nutrients besides nitrogen. Grasses usually have an advantage over legumes in obtaining nutrients, because grasses have a more fibrous root system. Legumes generally can't get enough nutrients until the needs of associated grasses have been satisfied. Fertilizer applications to pastures should therefore supplement soil nutrient levels to supply adequate amounts of all nutrients to meet the needs of both grasses and legumes.

Applying nitrogen, either in fertilizer or manure, right after the rapid surge of spring plant growth can minimize or prevent depression of grass production in midsummer. Autumn nitrogen application can help stockpile pasture forage for winter grazing. (see chapter 10)

But before you apply fertilizer nitrogen to increase plant growth, make certain that you are managing grazing properly so that your pasture plants are not being held back by some other aspect of grazing management. Make certain that your livestock are fully consuming the forage that's already being produced, and that they can eat the increased forage that will be produced. It's not uncommon for graziers to be simultaneously wasting forage and applying fertilizer to increase plant growth. This practice wastes money and resources.

PHOSPHORUS (P)

Besides nitrogen, phosphorus is the other most commonly deficient nutrient. Phosphorus contents range between 0.35 and 0.45 percent in pasture plants growing on soils containing adequate amounts of phosphorus. So a pasture producing 5 tons of dry forage per acre during a season must have available about 42 lb P/acre.

As with nitrogen, very little phosphorus leaves the pasture environment in livestock. For example, cow milk contains only 25 percent of the phosphorus eaten in forage, and a fat lamb contains only about 0.4 pound of phosphorus. Although little phosphorus is removed from pastures by livestock, pasture forage should contain an adequate amount of phosphorus for balanced animal nutrition.

In contrast to most other plant nutrients, phosphorus usually is adsorbed by soils and becomes only slightly soluble, so that losses from pastures by leaching are very small. So once an adequate level of soil phosphorus is reached, only small maintenance applications are needed because most of the available phosphorus continues to cycle within the pasture.

Grasses compete more strongly than legumes for phosphorus and, as discussed above, their competitive ability increases when high levels of nitrogen are present. Competition for phosphorus by grasses is an important inefficiency in grass-legume pastures that depend only on nitrogen fixed by the legumes. This fixed nitrogen is not exactly cost-free, because large amounts of phosphorus have to be used in meeting the needs of grasses before legume needs can be satisfied.

Most phosphorus excreted by grazing livestock is in dung. Of this, up to 80 percent is present in an inorganic form immediately available to plants. The rest of the phosphorus in dung is present in organic combinations that must be broken down by birds, insects, earthworms, and other soil organisms before it becomes available once again to plants. If white clover grows better where cow pies (dung pats) decomposed than elsewhere in the pasture, it may indicate that phosphorus level is too low in the soil.

Like nitrogen, phosphorus accumulates in livestock camps, transferring in from the rest of the pasture area. The same precautions and management that spread nitrogen more uniformly over paddocks, also distributes phosphorus more evenly.

If you need to apply phosphorus, you might consider using reactive rock phosphate; its slow phosphorus release can be more beneficial, compared to rapidly soluble fertilizer. It works best in soils with pH below 6, but good responses occur in soils with pH between 6 and 7 if there are signs of abundant insect, earthworm, and other soil organism activity. It is available from Texasgulf, PO Box 30321, Raleigh, NC 27622-0321, phone 919-881-2802.

POTASSIUM (K)
Luxury consumption doesn't just refer to affluent conspicuous consumers, but also to the gluttonous tendency of grasses to take up amounts of nutrients way in excess of their needs. This is especially true in the case of potassium. Because of luxury consumption, potassium content of grasses may rise to 5 percent of dry matter, although normal pasture forage contents range from 2.0 to 2.5 percent of dry matter. So without luxury consumption, a pasture producing 5 tons of dry forage per acre requires about 225 lb K/acre/year.

The amount of potassium actually removed from a pasture in livestock products is relatively small. For example, one milking cow removes only 9 pounds of potassium during the grazing season. Sooner or later though, this would deplete most of the potassium in a pasture, if more didn't continually become available from potassium-bearing soil material, fertilizer, or supplements fed to grazing animals. Potassium levels can become excessive in dairy pasture because of supplemental grain feeding.

Losses of potassium from the pasture environment can be very high, however, if dung and urine are distributed unevenly within the pasture. All excreted potassium is water-soluble; 90 percent of it is in urine, the rest in dung.

As with nitrogen and phosphorus, grazing animals collect potassium from the whole pasture or paddock and concentrate most of it within urine patches near or in stock camps. These small areas may represent only 15 to 20 percent of the total pasture or paddock area. The end result is a general decrease of potassium over most of the pasture or paddock, and excessive amounts of potassium in camps.

Within a urine patch, potassium may be applied at a rate of over 1,200 lb/acre! Even with luxury consumption by grasses, this amount can't all be taken up by plants and some is lost by leaching. An amount equivalent to 150 pounds of potassium per acre may be lost from one urine patch. This has serious consequences in that it depletes the pasture of potassium and may pollute ground water.

Plant growth responses within urine patches indicate helpful things about a pasture's nutrient status. These responses usually become apparent about 2 weeks after urine is deposited. First the grasses become darker green and grow more vigorously, as they are stimulated by nitrogen in the urine; this happens at the clover's expense, remember. Three to 4 months after the urine was deposited, the grasses use up the urine nitrogen. Once the nitrogen is gone, clovers respond to any potassium remaining in the urine patch, and reach their greatest growth within the patch at this time. As the potassium once again becomes limiting, clover content drops to its previous low level. So clover dominance in urine patches indicates potassium deficiency in a pasture or paddock.

If you need to apply potassium, consider using

sulpomag; it also provides sulfur and magnesium. Fall is the best time to apply potassium fertilizer, because plants need available potassium to harden for winter.

CALCIUM (Ca)
Plants require calcium in amounts nearly as large as those of nitrogen, phosphorus, and potassium. In fact, some plants contain more calcium than potassium. Calcium is important not only as a plant nutrient, but also in its relationship to soil acidity and structure.

Acid soils may contain too much aluminum, iron, and manganese, and too little calcium and magnesium. Applying liming materials (limestone and/or dolomite) adds calcium and/or magnesium, corrects soil acidity, and neutralizes byproducts of organic matter decomposition.

Although calcium is related to soil acidity, a soil's pH doesn't necessarily indicate the amount of calcium that's available; a separate test is needed to determine how much calcium is in a soil. Calcium content of a soil mainly depends on the parent material from which the soil developed. For example, calcium content may range from 0.5 percent in coastal plain sands, to more than 5 percent in soils of dry regions or in soils that developed from limestone, chalk, or marl.

Calcium improves soil structure by aggregating colloidal clay and humus particles, which makes soil more granular. Granular soil allows air and water to enter it easily, and this provides a favorable environment for plant roots and soil organisms to live and grow.

Calcium is involved in all sorts of processes in plants, including root and leaf development, nodulation and nitrogen fixation, cell elongation and division, protein synthesis, and water uptake.

Despite the many essential functions of calcium in pastures, calcium deficiency is rarely observed in the plants. This probably is because other problems, such as aluminum toxicity or phosphorus deficiency develop first on poor soils, and limit plant growth before calcium deficiency symptoms can be expressed. Grazing animals require large amounts of calcium, however, and may suffer from calcium deficiency if the plants they are eating contain low levels of calcium, even though the levels aren't low enough to limit plant growth. Large amounts of dandelions in swards may indicate that calcium level is too low in the soil. If soil particles stick to earthworms it indicates that soil calcium level is low; earthworms should be slimy and clean.

Soils generally contain relatively large amounts of calcium, and except for acid soils with pH levels below 6.0, usually enough calcium exists to meet the needs of growing plants. Some people have suggested that plant productivity and soil condition can be improved by making certain that calcium content in the soil reaches 85 percent of basic cation saturation. This is something that deserves more research. If you use this suggestion in determining calcium needs, make certain that the recommendation is based on *effective* cation exchange capacity (CEC), and not on one at a theoretical pH level of 7 or 8. This could save you a lot of money from not applying calcium unnecessarily.

Here's how to calculate calcium status of soil, using values from one of my soil tests for the example:
- Soil pH: 5.4
- Available calcium: 2420 lb/acre
- Effective CEC: 7.5 milliequivalents (meq)/100 grams soil
- 1 meq hydrogen/100 grams of soil is equivalent to 20 meq calcium/100 grams of soil, which is equivalent to 400 lb calcium/acre plow layer (6 inches)
- Divide amount of available calcium by calcium equivalent: 2420/400 = 6.1
- Divide the result by effective CEC and multiply by 100: 6.1/7.5 X 100 = 81% calcium base saturation.

New Zealand farmers have found that it's very beneficial to apply 200 to 500 lb limestone/acre/year to

pasture, regardless of soil pH, primarily for the added calcium. Large commercial applicators may not have equipment that can apply this small amount, but you might consider doing it with a lime spreader. This annual application of lime may be needed to neutralize acidity produced by grass residue decaying at the soil surface, and to obtain other benefits:
• Better germination of clover seed.
• Improved clover growth.
• More earthworms, which require high levels of available calcium.
• Less weeds that are favored by acidic conditions, such as buttercup.
• Improved movement and absorption of nitrogen, phosphorus, and magnesium.
• More abundant soil microorganisms.
• Better availability of elements such as molybdenum.
• Improved soil structure.
• Slower breakdown of soil organic matter and release of aluminum.
• Decreased grass pulling in new stands.
• Improved plant palatability.

When applying calcium usually the least expensive and most readily available material to use is finely ground limestone (calcium carbonate). If other nutrients are needed besides calcium, you can apply ordinary superphosphate (phosphorus, calcium, sulfur), dolomite (magnesium, calcium), or gypsum (sulfur, calcium).

MAGNESIUM (Mg)
Magnesium is intimately involved in photosynthesis, and chlorophyll contains about 2.7 percent magnesium. So if there's a shortage of magnesium for plants, photosynthesis will be limited and so will forage production. Because it's so important in chlorophyll formation, the first symptom of magnesium deficiency is

loss of healthy green color between leaf veins. The color gradually changes to yellow and then to reddish purple, as magnesium deficiency becomes more severe.

Besides being essential for plant growth, magnesium exists in every cell of animals and humans. The most common problem of magnesium deficiency in animals is called grass tetany, grass staggers, or hypomagnesaemia.

Magnesium content of soils varies a lot, depending on the parent material from which soils develop. Magnesium is most likely to be deficient for plants on sandy soils that have been limed with calcitic limestone and have received high rates of nitrogen and potassium fertilizers. Magnesium fertilizers include dolomitic limestone, magnesium oxide, magnesium sulfate (Epson salt), and potassium-magnesium-sulfate (sulpomag).

SULFUR (S)
The cycle of sulfur closely resembles that of nitrogen. Most sulfur in soils is present in an organic form, and only becomes available to plants after it becomes mineralized through microbial action. Since grasses are very competitive and may use 95 percent of any sulfur mineralized from soil organic matter, legumes must depend on other sources of sulfur.

The amount of nitrogen fixed by legumes in a given environment determines the amount of sulfur that they need from sources other than soil organic matter. If sulfur fertilizer isn't applied, the only other source of sulfur is from the atmosphere in rain and snow. The amount of sulfur required by legumes equals one twelfth of the nitrogen fixed. So white clover fixing 250 lb N/acre during a season, needs 21 lb S/acre! The amount of sulfur needed in a flourishing pasture is higher than what usually comes from the atmosphere in precipitation. Even acid rain may apply less than 10 lb S/acre each year in rural areas away from industrial centers.

Until recently, phosphorus fertilizer generally used

(ordinary superphosphate) contained about 12 percent sulfur. In concentrating phosphorus to about 45 percent of the fertilizer material (triple superphosphate), freight costs were lowered, but sulfur was eliminated. Consequently, in areas where only triple superphosphate is used, no sulfur is being applied to cropland, other than what comes down in rain or snow. Of course, permanent pastures in the United States usually haven't been considered worthy of fertilization, so they weren't intentionally fertilized with any nutrients, including sulfur! Sometimes manure that contained sulfur was spread on permanent pastures, but this mainly benefited pasture grasses. So sulfur deficiency may be limiting plant growth in pastures, especially permanent pastures.

Many soil- or tissue-testing labs don't analyze for sulfur, because the analysis is more difficult and time-consuming than for other elements. One way to find out if sulfur levels are too low in your pasture is to spread 100 lb S/acre (to be certain enough is applied) as gypsum (calcium sulfate) evenly over a few measured areas of your pasture. For example, measure off 5 by 20 feet (100 square feet) areas in different locations. Since gypsum contains about 15 percent sulfur, you will need to spread 1.5 pounds of gypsum over each 5- by 20-foot area, to apply 100 lb S/acre to that area. Assuming that other nutrients are present in adequate amounts (and you can determine this by soil testing and following the recommendations), within 1 to 2 months plants should begin flourishing in the sulfur-fertilized areas. If this occurs, the pasture needs sulfur, and you should apply at least 25 lb S/acre/year.

Probably the best way to apply sulfur is along with phosphorus in ordinary superphosphate, or along with potassium and magnesium in sulpomag (potassium-magnesium-sulfate). If you don't need to apply other nutrients besides sulfur, use gypsum. Don't apply elemental sulfur (yellow stuff that smells like rotten-eggs), because sulfuric acid results from the microbial transformation of elemental sulfur to sulfate that occurs in soil, and this lowers soil pH.

BORON (B)

Boron was first shown to be a deficient nutrient in Vermont. Alfalfa and the clovers are very sensitive to low levels of boron in soils. A symptom of boron deficiency is shortened, rosette-shaped plants. Leaves turn yellow and look as if they have been damaged by drought. Boron deficiency symptoms are most likely to develop during dry spells and after applying large amounts of lime. For this reason, lime applications in Vermont are limited to 2 tons per acre per year.

No more than 2 lb B/acre should be applied every 3 years to grass-clover pastures. Although the legumes may need more boron than this, higher rates might kill associated grasses. If boron is deficient in your area, follow the recommendation of your soil testing laboratory in amounts and times of application. Be very careful in applying boron; too much can easily kill grasses. It's best to test the boron application in 5- by 20-foot test areas or in a strip across the pasture, to make certain that the rate you apply won't kill grasses, before applying it to the entire pasture.

MOLYBDENUM

Research in Vermont showed that molybdenum may be nearly deficient in many soils of the state. We have gotten positive responses from applying molybdenum in combination with phosphorus and lime, on certain soils in which phosphorus and lime application alone had not improved plant growth very much. If you are managing the grazing properly and have paid attention to all other nutrients, and your pasture plants still don't grow well after 2 or 3 years, molybdenum may be lacking. It is involved in nitrogen fixation, so if it's deficient, little nitrogen will be fixed.

Molybdenum is more available in alkaline soils and in soils having high organic matter contents. Liming soils makes molybdenum more available; in fact, response to liming may be due to better molybdenum availability in some soils.

Be certain that your soils actually need molybdenum before applying any, by having soils and plant tissues analyzed. Try liming the soil to raise the pH and make molybdenum more available before applying it. Always get a second opinion on any recommendation that involves applying molybdenum. Be extremely careful if you apply molybdenum! No more than 0.18 pound (80 grams) should be applied per acre, because it's easy to reach toxic levels of molybdenum. That amount can't be spread uniformly unless it's contained in a larger quantity of another fertilizer, such as ordinary superphosphate. As with sulfur and boron, it's best to apply molybdenum in test areas to see what happens before applying it to the rest of the pasture.

CHLORINE, COBALT, COPPER, IODINE, IRON, MANGANESE, NICKEL, SELENIUM, SILICON, SODIUM, ZINC

These elements seem to be present in adequate amounts for plant growth in most pasture soils. If your pasture plants don't grow well after you have attended to all of the other elements, and you're using proper grazing management long enough (2 to 3 years) for soil organisms to develop to adequate levels, it is possible that one of these may be deficient. Sandy (e.g. central sands area of Wisconsin) and eroded soils may be deficient in zinc. Soils in northeastern North America generally are deficient in selenium, but currently this element is available only in livestock mineral mixes so you can't apply it as fertilizer.

Be careful that you don't apply these or any other nutrient unnecessarily. The best way to determine if you need any nutrient, including these, is by having soils and plant tissues analyzed and recommendations made by a lab that doesn't sell fertilizers of any kind. It's always best to get a second opinion on any fertilizer recommendation that involves applying these elements. Test any application of these elements in small areas before applying them to the rest of your pasture. If you apply too much of any of these or other nutrients such as molybdenum or boron to pastures, it's too late to avoid serious problems.

SOIL ACIDITY

Technically, soil pH is a measure of hydrogen ion activity in soil solution. Practically, it's an indicator of one aspect of soil that influences almost everything else in soil.

It's difficult to determine the relationship between composition of a pasture sward and soil pH, because pH is intimately related to so many things in soil, and because it varies a lot in different layers of pasture soils. The pH value you get in a soil test is an average measurement of acidity over soil sample depth, but pH of the top 1 to 1.5 inch of pasture soil is known to be higher than that of the 4-inch layer underneath. Probably quite different nutritional and soil organism relationships exist in the top layer, compared to the layer beneath it. Also, soil of a permanent pasture usually has a lower pH than the same soil type under cultivation. The pH layers and lower pH of pasture soils need more research and accommodation by soil testing labs so that the information can be applied.

Soil pH directly or indirectly affects soil organisms, all plant nutrients, and plant growth. Plants that are well adapted to prevailing conditions grow best and make up the largest part of pasture swards.

Soils become acidic when large portions of the exchangeable cations are hydrogen and hydrated aluminum. Some acidic soils develop from acidic parent materials, but most develop from the process of leaching. As water that contains hydrogen cations from weak acids (e.g. organic and carbonic acids) moves through soil, some adsorbed exchangeable cations (e.g. cations of Ca, Mg, K, and Na) are replaced by the hydrogen. These cations are then carried by the moving water below the rooting zone or into the groundwater. In the process soils become more acidic and contain less of required plant nutrients.

As soil pH decreases below 5.5 to 5.8, bacteria become less active. This directly influences rates of nitrogen fixation and transformations, organic matter decomposition, and the sulfur cycle. The result is that nitrogen and sulfur are less available to plants. Calcium, potassium, magnesium, molybdenum, and phosphorus also are less available to plants in soils with pH lower than about 6.0. Boron is less available in soils when pH is below 5.0. Cobalt (required by microorganisms, but apparently not by plants), copper, iron, manganese, and zinc all become more available as soil pH decreases below 6.5. Aluminum, which is toxic to most plants, is much more active and influential below pH 5.5. In very acid soils available aluminum and manganese can be present in toxic levels.

As soil pH increases, the reverse of all of the above happens. So it's clear that soil pH doesn't exist in a world by itself: every change in pH brings about many changes in other aspects of the soil. And vice versa; it's another situation where we can see that everything is interrelated.

The question then comes up of when is it necessary to apply lime to a permanent pasture, and what materials to use? New Zealand farmers have found that it's very beneficial to apply 200 to 500 lb limestone/acre/year to pasture, regardless of soil pH, primarily for the added calcium (see list of possible benefits in the calcium discussion above). Liming materials are usually the carbonates, oxides, hydroxides, and silicates of calcium and magnesium. Choose a liming material based on its cost in relation to purity, ease of handling, availability, and speed of its reaction with the soil. Use finely ground material so that the reaction occurs quickly.

Generally it's best for pasture plant growth and productivity if soil pH is maintained in the range of 5.5 to 6.5. Because of acids released and formed in the soil during the grazing season, soil pH can cycle down one unit from spring to fall. So to measure soil pH at its lowest level, it's best to take soil samples and have them analyzed in autumn. Also apply lime in autumn to allow time for it to begin reacting and raising pH by spring.

Avoid spreading too much lime at any one time; it's best to spread a little often, rather than a lot infrequently. For example, spreading 500 to 1000 lb of finely ground liming material per acre per year for 4 years is better than spreading 1 to 2 tons all at once.

SPREADING MANURE ON PASTURE

As the amount of land on farms devoted to corn and other row-crop production decreases in favor of perennial forages for grazing, more manure collected in the barn is being spread on pastureland. In the long term this can only improve pasture productivity by increasing soil organic matter and fertility levels. Care must be taken in the short term so that manure application doesn't damage pasture swards and decrease forage use by repelling grazing livestock. Manure spreading on pasture needs to be done within a management intensive grazing context for best results of maintaining white clover in the sward.

Most feed nutrients pass through animals and ends up in manure. More than 70 percent of the nitrogen, 60 percent of the phosphorus, and 80 percent of the potassium in feed can be present in manure for recycling onto pasture and other cropland. Besides these nutrients, manure contains large amounts of calcium, magnesium, sulfur, and trace elements. Manuring also increases soil organic matter content, which increases soil cation exchange capacity and raises pH. All of these benefits from manuring can combine to reduce the need of applying fertilizer and lime.

Farmers' experience in Vermont has shown that spreading liquid manure from a storage pit on pasture is beneficial. Spreading 3000 to 4000 gallons/acre right after paddocks have been grazed during mid-June to mid-July, and again after the final grazing in the fall works well. It's best if midseason applications are followed by rainfall so that livestock will graze the forage well in the next rotation.

The midseason application speeds regrowth of pasture plants, probably due to the water and available nitrogen in the manure. On one farm, plant regrowth after liquid manure application is so fast that only 10 to 12 days recovery time are needed between grazings in midsummer, while other paddocks not receiving liquid manure need more than 20 days recovery time.

Dairy cows grazing after a liquid manure application tend to graze less close to the soil surface than before, but graze more uniformly. It's possible that the generally pervasive smell of liquid manure overrides the repugnant smell of individual cow pies. Cows then reject less forage around cow pies, thereby improving use of the total forage produced.

One other benefit of liquid manuring seems to be the control of Canada thistle (*Cirsium arvense*). Probably because of the salts contained in manure, thistles covered with liquid manure dry out, die, and disappear.

Although any kind of manure application to pasture improves soil nutrient and organic matter status, spreading fresh manure that is more solid than liquid does not have the above beneficial effects on grazing and weed control. Unless it is pulverized by the spreader, chunks of manure and bedding smother plants, create surrounding zones of repugnance and rejected forage, and result in animals refusing to graze close to the soil surface. Applying composted manure to pasture probably wouldn't repel grazing livestock, but needs to be pulverized to avoid smothering plants.

REFERENCES

Cramer, C., G. DeVault, M. Brusko, F. Zahradnik, and L.J. Ayers. 1985. *The Farmer's Fertilizer Handbook*. Regenerative Agriculture Assoc., Emmaus, Pennsylvania.

Dalrymple, R.L. 1994. Nutrient recycling. *Stockman Grass Farmer*. 51(8):10.

Donahue, R.L., R.W. Miller, and J.C. Shickluna. 1983. *An Introduction to Soils and Plant Growth*. Prentice-Hall, New Jersey.

Follett, R.H., L.S. Murphy, and R.L. Donahue. 1981. *Fertilizers and Soil Amendments*. Prentice-Hall, New Jersey.

Hecksel, J. 1993. Fertilizer rules of thumb. *Stockman Grass Farmer*, 50(11):2, 26.

Jones, U.S. 1982. *Fertilizers and Soil Fertility*. Reston. Virginia.

Jones, V. 1991. Liming pastures has multiple benefits. *Stockman Grass Farmer*. 48(2):15.

Jones, V. 1995. Sampling pasture. *Stockman Grass Farmer*. 52(1):21.

Jones, V. 1996. How to read your pasture plants. *Stockman Grass Farmer*. 53(7):21.

Magdoff, F. 1992. *Building Soils for Better Crops*. University of Nebraska Press. Lincoln. 176 p.

Miller, D.A., and G.H. Heichel. 1995. Nutrient metabolism and nitrogen fixation. p. 45-53. In R.F. Barnes, D.A. Miller, and C.J. Nelson (eds.) *Forages, Vol I*. Iowa State University Press.

Martyn, R. 1994. Soil pH and reactive phosphate rock. *Stockman Grass Farmer*. 51(7):28-29.

Nation, A. 1993. Production per man more important than per cow. *Stockman Grass Farmer*, 50(3):1-10.

Nation, A. 1994. Reactive rock phosphate source. *Stockman Grass Farmer*. 51(4):23.

Nation, A. 1995. Quality pasture - part II. *Stockman Grass Farmer.* (1):13-16.

Salatin, J. 1993. One day of cow manure: what's it worth? *Stockman Grass Farmer.* 50(9):11-12.

Scharabok, K. 1992. Calcium - one of the most cost effective soil treatments.*Stockman Grass Farmer.* 49(8):16-17.

Scott, W.R. 1973. Pasture plant nutrition and nutrient cycling. p. 159-178. In R.H.M. Langer (ed) *Pastures and Pasture Plants.* A.H. & A.W. Reed, Wellington, New Zealand.

Semple, A.T. 1970. *Grassland Improvement.* Leonard Hill Books, London. 400 p.

Smith, B., P. Leung, and G. Love. 1986. *Intensive Grazing Management.* The Graziers Hui, Kamuela, Hawaii. 350 p.

Vallis, I. 1978. Nitrogen relationships in grass-legume mixtures. p. 190-201. In J.R. Wilson (ed) *Plant Relations in Pastures.* CSIRO, Melbourne, Australia.

Voisin, A. 1959. *Grass Productivity.* Philosophical Library, New York.

Voisin, A. 1960. *Better Grassland Sward.* Crosby Lockwood, London.

Watts, A. 1972. *The Book: On the Taboo Against Knowing Who You Are.* Vintage Books, New York. 151 p.

4
Grazing Animals: Effects On Pastures & Vice Versa

The wild geese do not intend
To cast their reflection;
The water has no mind
To receive their image.

Zenrin

Grazing animals can make or break a pasture by causing very rapid and large changes in plant productivity and botanical composition of the pasture sward. These changes result from removal of plant leaves (defoliation), excretion of dung and urine by grazing animals, treading action of animals' hooves, and seeds dispersed by animals. Let's look at each of these separately to understand how they influence a pasture.

DEFOLIATION
Defoliation probably is the most important effect that grazing animals have on pasture. We saw in chapter 2 how important light relationships are in a pasture sward in determining which plants will be dominant and which will be present in different numbers. Grazing drastically decreases leaf area of plants. This reduction of a plant's leaf area affects its carbohydrate use and storage, tiller and stolon development, and leaf and root growth. Defoliation also changes a plant's microenvironment of sunlight intensity and soil temperature and moisture.

Plants differ in their physiological and growth-form responses to defoliation, and in their growth rhythms during the season. Because of these differences, grazing can change the relative abundance of the various plants in a pasture, depending on which plant species are best adapted to weather conditions after grazing. So dominant species and the composition of a sward can change several times during a season, as weather and light conditions change and are modified by grazing. If we made a time-lapse movie of these botanical changes, we would see them as waves passing through the sward, reflecting environmental and grazing influences.

When all or most of the leaves are grazed from a plant, root growth stops and the plant uses its carbohydrate reserves in roots and stubble for the energy it needs to grow new leaves and stems. As carbohydrates are withdrawn from roots, they die and separate from the plant (see Plant Roots in chapter 2). The plant draws on reserves until enough leaf surface develops for photosynthesis to be adequate to meet energy needs for further leaf and root regrowth. Grasses depend on reserves for 2 to 7 days after grazing. This can take longer in other plants such as alfalfa, which needs about 21 days to form enough leaf surface for photosynthesis to provide adequate energy for subsequent growth.

Plant response and survival in a pasture depends greatly on use of reserves for regrowth, and weather conditions after grazing. Low temperatures and/or dry conditions slow regrowth of defoliated plants for long periods, during which the plants continue to draw on carbohydrate reserves. Continuous grazing, by repeatedly removing leaves and reducing reserves during and after adverse weather, decreases survival and regrowth of stressed plants. Development of cold hardiness, especially for alfalfa, may be lessened either by complete defoliation during the 6-week hardening period of September to mid-October, or by grazing too many times during the season.

Grazing all or most leaves from a plant drastically affects its ability to take up nutrients and water from soil. If root growth stops or slows because of defoliation or any other reason, uptake of water and nutrients from soil is directly limited. When leaves are removed in grazing, flow of water from the soil through the plant to the atmosphere (evapotranspiration) decreases, and this also reduces uptake of nutrients. Removal of leaves also decreases the amount of carbohydrates being produced in photosynthesis and their movement to roots. In fact, carbohydrates may begin to flow entirely from roots to shoots as the plant regrows. During these times water and nutrient uptake decrease and the plant may be stressed even more.

Yin and Yang
Like everything else, defoliation of pasture plants has two sides to it. Reasonable amounts of defoliation by grazing causes more branching and lower growth forms of plants. These result in a tighter sod, more covering of soil against erosion, and higher yields of more nutritious forage. Also, plants having more branches or stolons, and less shading of branches or stolons, tend to have more flowers and produce more seeds. This tendency may be very important in maintaining a high percentage of red and white clovers and birdsfoot trefoil in intensively grazed pastures.

Another aspect of this other side of defoliation, is that the removal of nutrients from plants by grazing their leaves and stems results in movement of nutrients from the soil root zone to plant tops. In this way, defoliation and digestion by grazing animals is an essential part of rapid cycling, return, and transfer of nutrients in a pasture environment. Uneaten plants or plant parts, in contrast, slow nutrient cycling because nutrients are unavailable until the plant material breaks down through weathering in the air or biological decomposition on the soil surface. Controlled high stocking densities reduce the amount of uneaten plant material, and through hoof action help break down uneaten material so that it decomposes faster, thereby quickening nutrient cycling.

Root pulsing is another aspect of defoliation and regrowth that we're just beginning to appreciate. Every time roots separate from plants due to removal of carbohydrates to regrow plant tops, organic matter and carbon are added to the soil. As roots regrow and reestablish, more carbon is removed from the atmosphere through photosynthesis, thereby helping to lower the atmospheric carbon dioxide level that may be causing the Earth to become warmer. Increasing the total land area in grassland and improving grazing management could reduce atmospheric carbon dioxide to a safe level (see chapter 11).

SELECTIVE GRAZING
Animals don't graze uniformly. That's why it's so difficult to have animals graze high to favor some plants. Animals select certain parts of a plant or particular plant species while ignoring others, if allowed to do so. This results in uneven grazing of pastures, which stresses individual plants differently in the sward, because of the variable vertical distribution of leaf, stem, floral, and growing tissues in the plants. Even when animals graze down to a relatively uniform height from the soil surface, some plants are still favored over others because of this variable vertical distribution.

Given the chance, all grazing animals are selective in their diet. Within a pasture, grazing animals generally have many choices of what to eat from a range of plant species with varying proportions of leaf parts, stem, flowers, and seed. Sheep and cattle tend to select leaf parts in preference to stem, and young leaves before old leaves, especially when pasture forage has become too mature. The plant material that animals select usually is higher in protein, phosphorus, soluble carbohydrates, digestibility, and energy, and lower in lignin and structural carbohydrates (fiber), compared to the total forage available in a pasture.

Animals know instinctively what they like and usually need. It seems that their selection of one plant species or plant part in preference to others depends on what we call palatability, accessibility, and availability.

Some plants or plant parts taste or smell badly or have picky structures that irritate animals' mouths. These kinds of things decrease palatability of plants and discourage animals from eating them. Some plants or plant parts aren't accessible to animals. Their mouth formations may not allow them to eat a plant, a plant may grow too low to the ground, or a plant part may be protected by thorns.

As pasture forage availability decreases, selectivity also decreases; forage that wasn't acceptable before, now will be eaten. Animals adjust, so that, while continuing to look for preferred plants or plant parts that are less accessible, they eat more plants or parts that they had previously rejected. In this stage animals take longer to graze because they spend more time searching for forage they prefer. They also take more bites, because their bites become smaller.

As stocking density increases, differences in relative acceptability among plants or plant parts practically disappear. Animals will reduce intake, though, when left with greatly disliked plants or plant parts to eat. So be careful to provide enough acceptable forage for high-producing livestock, especially in early stages of pasture improvement.

Animal grazing selectivity is exactly the same thing that happens when children have a choice between eating lettuce or spinach: almost invariably they eat the lettuce and leave the spinach. If, however, they would be given the choice of eating both lettuce and spinach or starving, they would eat both. A grazing management method that gives animals the choice of eating all available pasture forage or starving, puts you in control of the situation and allows you to influence the botanical composition so that it benefits animals and pasture. You will find that most animals will choose to eat the forage before going hungry.

For all of the above reasons, different botanical compositions of pasture swards result under different grazing management practices. Close, continuous grazing with a high stocking density leads to a sward containing species that tend to be low-growing, with rhizomatous, stoloniferous, or basal rosette growth habits. Extensive (lax) continuous grazing with a low stocking density results in a sward rich in plant species, but these tend to be mostly tall-growers such as milkweed, goldenrod, burdock, buttercup, and thistles. Rotational grazing with a low level of stocking density and forage use also results in a sward dominated by tall-growing plants. Management intensive grazing, with high stocking density and forage consumption levels, usually results in simpler pasture mixtures of one or two legume species, several grass species, and two or three forbs (dandelion, chicory, plantain), which complement each other and are well adapted to the management.

If water and nutrients are not limiting, changes in a pasture sward caused by grazing mainly reflect grazing frequency and intensity. The changes result from defoliation, depending on whether and for how long sunlight reaches levels in the canopy where low-growing plants (e.g. white clover) have their leaves. Arrangement of leaves and stems in a sward not only affects sunlight penetration into the plant canopy and selective grazing, but also influences grazing time and forage intake.

GRAZING BEHAVIOR AND FORAGE INTAKE

A lot of research has been done on mechanical harvesting and storage of livestock feed, but very little on animal behavior and how animals harvest their own forage while grazing pastures. Grazing is a dynamic interaction between animal, plant community, soil, and humans. If we understood more about how animals harvest their forage, we might be able to help them to eat more, and we could get better forage and livestock yields from our pastures. Although most existing information on grazing behavior is for cattle, a look at how cattle graze also can indicate how other kinds of livestock might graze.

Grazing time involves the entire process of moving around searching for food and then browsing (actually eating) the forage that is found. Grazing time in cattle almost never lasts more than 8 hours per 24-hour period. Browsing itself lasts less than 5 of the 8 grazing hours per day. This is an extremely important point: grazing time is the same regardless of pasture quality or amount of forage available. During the 8-hour grazing times, cows make all the effort that they can, and they can't graze any longer without resting. If you consider that a cow's mouth is only about 3 inches wide and she needs to eat about 150 pounds of green forage each day, it's understandable why a cow would be tired after working for 8 hours to harvest forage. Obviously, if we can help animals to harvest the forage they need, while using as little energy as possible, livestock productivity will improve.

About 60 percent of cattle grazing occurs in daylight, and 40 percent occurs at night. (Very little if any sheep grazing occurs at night.) As air temperature rises, the proportion of grazing done at night increases.

During 1992 and 1993 we studied daytime grazing behavior of high-producing Holsteins on a Vermont dairy farm. Cows were given a fresh paddock after each milking, usually entering the paddock about 9 a.m. and returning to the barn for milking at 4 p.m. At night, cows began grazing a fresh paddock about 8 p.m., and returned to the barn at 5 a.m. So cows actually were in paddocks 7 hours during the day, and 8 hours at night. We found that most (85%) of the cow's total daytime grazing occurred between 9 a.m. and 1 p.m. Only about 15% of grazing occurred from 2 to 4 p.m., and actually most cows started moving toward the paddock gate after 2 p.m., probably anticipating going to the barn to be milked and fed supplements. By 3 p.m. 87% of the herd gathered near the gate, and were either resting or ruminating.

Grazing animals walk as much as 2.5 miles per day, depending on forage availability. Eighty percent of the walking occurs during daylight, which corresponds to the fact that most grazing activity happens during the daytime. Probably grazing at night is more efficient, because animals are pestered less by flies and other insects at night, and consequently walk around less.

While grazing, cows move forward swinging their heads from side to side within an arc of 60 to 90 degrees, and taking 30 to 90 bites per minute, if the forage is the right length. Forage length has an important effect on a cow's rhythm of eating. If a cow is grazing very long forage (10 to 14 inches), she either eats the upper 2.5- to 3-inch layer, or she tears off a mouthful about 12 inches long. If she takes in the long forage, she can't swallow such a large, long mass without chewing it first, and chewing it requires about 30 seconds per mouthful. In comparison, a cow grazing forage that is only about 6 inches tall, can swallow 30 mouthfuls in 30 seconds. The best forage height for sheep and goat grazing is about 4 inches tall. So animals grazing short forage can eat more during a day than when they graze long forage.

The amount of time that cows are capable of grazing and the number of jaw movements they can make per day apparently are inherited characteristics. In one study, for example, a heifer made 29,150 jaw movements per day, compared to only 18,781 jaw movements in another heifer. The first heifer made 55% more jaw movements and grazed 46% longer than the second heifer. Animals that are capable of longer grazing times and more jaw movements can eat more than other animals that are limited in the effort they can make. Selective breeding of animals for efficient use of pasture forage could result in less protein and energy supplements needed for animals grazing well managed pastures.

Cows ruminate about 8 to 10 hours per 24-hour period.

Part of the ruminating occurs while cows are lying down, and part while standing up. Cows lie down for about 12 hours a day, and this total resting time usually is divided into nine unequal periods of 1 to 6 hours.

GRAZING HABITS

On top of all this, different kinds of animals graze differently. If permitted, sheep and goats graze more selectively than cattle, because the large jaws and grazing action of cattle don't allow them to select as precisely as sheep and goats can. Also, milking goats have higher nutritional needs than milking cattle or sheep, and this causes goats to graze more selectively.

Since they only have front teeth on their lower jaws, ruminants grasp plants with their tongues and/or pinch forage between their lower teeth and upper mouth pads as they sweep their heads back and forth. Because of the position of their teeth and muscles of the lower jaw, cattle can't graze closer than 1/2 inch from the soil surface. Cattle can eat relatively mature forage, as long as the plants are not coarse or prickly.

Sheep and goats have mouths that enable them to cut plants off at the soil surface, taking away plant parts needed for regrowth. Unmanaged sheep and goats may even tear up entire plants on overgrazed pasture. Although sheep and goats require close management to prevent them from damaging pastures, a well managed sheep or goat flock can improve a pasture quicker and better than cattle can. This is because sheep and goats will eat everything under high stocking density, including young shoots of bushes, and don't have as much of a problem of avoiding forage around their dung droppings, as cattle do around their cow pies.

Horses also can grip plants and cut them off closer to the ground than cattle can. Horses graze very selectively, making it difficult to get a good botanical composition in a pasture grazed only by horses. So it's best to follow horses with sheep or cattle, or graze them all together, to clean up what the horses leave. Otherwise you should mow each paddock at least once or twice per season to cut what the horses didn't eat, and keep the sward in good condition for subsequent grazing.

All livestock can be put out to pasture. All can be managed in similar ways, as long as characteristics of that kind of livestock are taken into consideration. Pigs, for example, usually must have their snouts ringed before allowing them out. Otherwise they may dig up the soil and ruin the pasture, especially if you try to get them to eat everything in each paddock. If pigs are ringed and their grazing is managed, a good quality botanical composition results. Geese and ducks tend to tear plants when eating, and their grazing favors development of plantain, sorrels, and smartweeds, because they prefer grasses to broadleaf plants. If confined too long in one area, chickens eat and dig up all of the plants that are present.

TREADING/TRAMPLING
Treading or trampling by animals' hooves is another aspect of grazing that influences and changes pasture botanical composition, because of its direct effects on plants and indirectly through its effects on soil and soil organisms. All livestock can cause treading damage, particularly when grazing is not managed. The potential seriousness of this can be appreciated if you consider that an animal makes 8,000 to 10,000 hoof prints per day. If each hoof print covers 14 square inches as in cattle, the total area stepped on is 0.02 acre per day per animal! Horse hooves, especially when shod, may do a lot of damage to soil and plants that can't withstand treading.

The affect of treading directly on pasture plants depends only slightly on soil type and fertility and plant height, but is strongly influenced by plant species and soil moisture. Plant species differ in resistance to treading. Orchardgrass, perennial ryegrass, and white clover are very resistant to treading, compared to birdsfoot trefoil, red clover, and timothy, which are more sensitive. As stocking density increases, plants sensitive to treading decrease in number, and the botanical composition shifts to those plants that can resist the amount of treading that occurs.

Soil Compaction and Puddling

Treading indirectly affects pasture plants by compacting (increase in density) and puddling (reduction in air space) the soil, especially if the soil is moist and fine-textured. Under moist soil conditions compaction and puddling result in increased mechanical resistance to plant root extension through the soil, thereby limiting water and nutrient uptake. They also reduce soil aeration and oxygen content, which slows root growth and metabolism, and limits nitrogen fixation. Temperature relationships also change when soils are compacted or puddled, and are reflected in above-ground growth in one way or another. Moisture availability changes in ways that may be favorable or unfavorable. Soil compaction and puddling decrease water infiltration and increase runoff, which leaves less water available to plants and erodes the soil. When soils are wet, plants and soil can be severely damaged by deep treading or pugging (poaching). Any of these things may depress plant growth, depending on how well the plants can adapt to the changed soil conditions. (see discussion of wet weather grazing management in chapter 5)

Depending on animal species, treading affects on soil differ. On a Vermont fine sandy loam, for example, sheep grazing compacted the soil less than Holstein heifers, under a moderate stocking density of 80 animal units/acre/24 hours (1 animal unit = 1,000 lb liveweight). Apparently because sheep weigh less per animal than cattle, sheep exert less pressure per square inch on soil.

Since plants, animals, and soils evolved together, treading has always been a part of their experience. Pasture plants, especially grasses, need to be grazed and probably treaded or trampled to avoid clumped dead growth that would shade growing points and hold nutrients from recycling. Under natural conditions, positive aspects of treading or trampling must have been more pronounced than negative effects. Otherwise, the huge grassland expanses of the world would not have developed under grazing by the enormous number of animals that lived in those regions. Fencing, domestic livestock, and poor grazing management changed all that, and detrimental effects of treading on plants and soil became more evident.

Under management intensive grazing, livestock are concentrated and limited in the amount of time they spend in a paddock. This in effect simulates natural grazing with its quick, concentrated grazing, adequate recovery, and more positive treading effects on pasture plants and soils. For example, when dairy cows graze each paddock for 12 hours and rotate through the paddocks six times during a season, they only spend 3 days total time in each paddock per year. That leaves 362 days for plants, soil, and soil organisms to recover from grazing and treading, and make use of nutrients that the cows left behind. This differs greatly from the effects of unmanaged continuous grazing where livestock continually affect plants, soil, and soil organisms.

(In a brittle environment, controlled trampling or herd effect can be beneficial in breaking soil crust, to improve water infiltration and seedling establishment. It also breaks down uneaten plant material, enabling decomposition to proceed more rapidly. See *Holistic Resource Management* by Allan Savory in References.)

EXCREMENT
Cattle poop 11 to 12 times and pee 8 to 11 times each day. When combined with grazing behavior, cattle and horse excretion especially can badly affect a pasture. When

grazing is properly managed, dung and urine greatly benefit a pasture by returning nutrients, increasing soil organic matter, and favoring the development of earthworms and other soil life. Let's look at each part of excrement to understand the effects.

Dung

The daily amount of poop produced per cow can weigh 50 pounds or more. Horses are close behind with 40 pounds per animal per day. This means that during a 180-day grazing season, 9000 pounds of dung are deposited per cow, and 7200 pounds per horse! If each cow pie measures 10 inches in diameter, for example, the total area covered by one cow (assuming no overlap of pies) would be 7 square feet per day, or 1,260 square feet during the grazing season! Loose dung from cows grazing lush pasture doesn't affect plants very much because it spreads thinly and decomposes quickly. But horse turds and cow pies from animals grazing mature forage or receiving grain supplements are drier, and cause major changes by blocking sunlight and killing most plants that are directly under the dung. The area in which the plants were killed can then be invaded by surrounding plants or from seeds that were in the soil or dung. Sheep pelletize their dung, so it's spread uniformly and breaks down rapidly when their grazing is managed properly.

When a cow pie hits the ground, it immediately has a "zone of repugnance" around it that measures about 25 feet in diameter! Consequently, at low stocking densities a lot of forage can be rejected around cow pies. High stocking densities can decrease the zone of repugnance down to the dung itself, but it isn't a good practice to force animals to eat right up to their dung because of the parasites it contains, and because dry matter intake probably would be decreased.

Horses deposit their dung in the same place and don't eat in that area. This causes forage in that area to become much lower in quality, unless it is mowed or grazed by other animals. Horse grazing and excreting habits especially, reflect their instinctive way of avoiding parasites present in their dung, so they shouldn't be forced to eat plants growing among their dung deposits. It is far better to graze that forage with other animals or mow and/or harrow it the same day that the horses are removed from the paddock.

At first, the dung itself causes animals to reject the herbage near it, probably because of unpleasant odor, but later the forage becomes too mature and then is unpalatable because of its coarseness. In poorly managed pasture, herbage around dung patches may be rejected for as long as 18 months. In well managed pasture soil life becomes enlivened, and dung can be disintegrated and incorporated with the soil by earthworms and insects in about 60 days, and the zone of repugnance disappears.

During the grazing season each cow deposits about 38, 8, and 8 pounds of nitrogen, phosphorus, and potassium in dung somewhere in the pasture. During the same time, each horse drops 40, 10, and 24 pounds of the same nutrients in dung, usually in very limited areas of the pasture or paddock. These nutrients are worth a lot of money and can be extremely beneficial to the pasture and your bank account, but not when the dung is stacked up. It must be spread around. Besides releasing nutrients, spreading exposes more of the dung to sunlight and drying, which kills parasites and reduces breeding sites for flies. How it's spread is up to you.

Under management intensive grazing with cattle, the high stocking densities and short grazing periods of paddocks usually result in more uniform grazing and distribution of dung, and breakdown of dung by hoof action of the densely stocked animals. Mowing and harrowing usually are unnecessary. For example, we obtained no benefit from harrowing (mowing wasn't needed) during a 2-year experiment on a Vermont dairy farm with leader-follower grazing of a Kentucky bluegrass-orchardgrass-white clover sward. In the experiment, pregrazing mass was 2400 lb DM/acre (6 inches tall) and postgrazing mass was 1200 or 1000 lb DM/acre (1 or 2 inches tall). Lactating cows grazed first, down to 1400 lb DM/acre; heifers followed, grazing down to the residual masses.

In New Zealand, where grazing is planned and managed at a level beyond anything done yet in the USA, mowing or harrowing routinely after cattle to clip uneaten forage or spread dung isn't done because it's unnecessary. Cattle grazing lush pasture tend to have loose dung, which spreads out thinly when dropped, and decomposes quickly. Pastures are harrowed sometimes during the dry season, however, when cattle graze forage higher in dry matter; this results in drier dung that doesn't break down easily. So it's possible with management to eliminate or minimize the need, and therefore the expense, labor, and fuel consumption of mowing and harrowing to remove uneaten forage and spread dung after cattle.

Grazing only one animal specie always tends to shift the botanical composition away from the preferred forage of that animal, because of rejection around its dung. You can maintain high-quality pasture (especially on rough land that can't be mown) by grazing more than one kind of animal, such as grazing cattle, horses, sheep, goats, pigs, chickens, or turkeys together or one group behind the other. For example, in Ireland a diverse leader-follower practice results in excellent pasture: lactating cows graze first, baby calves second, followed by 2-year-old heifers and sheep together. Cows and calves are moved twice a day, others are moved once a day.

You can also encourage birds to feed in your pasture by putting up birdhouses for them to nest in. Birds digging and scratching in dung for insects help break it down and spread it around.

Horses are a special problem because high stocking density doesn't work very well with them. Three possible ways of keeping horse pasture forage in good condition and breaking down and spreading horse dung are:
• Follow horses through paddocks with heavily stocked cattle or sheep.
• Mow paddocks with a rotary mower.
• Drag paddocks with a flexible tine harrow.

If the pastureland is level to rolling, you can use machinery to mow uneaten forage and spread dung if needed during the early stages of pasture development. Using a flail or rotary mower breaks up and spreads dung at the same time. If you use a sickle mower or don't mow, it may be beneficial to drag paddocks with a flexible tine harrow to break up and spread dung. If you harrow to spread dung, do it in midsummer when conditions are dry, to kill parasite larvae. If you harrow during moist spring conditions, you'll spread parasite eggs and larvae, contaminating more pasture forage.

Urine

The area of pasture affected by urine varies with soil moisture conditions, slope, and type. Urine spreads further in wetter soil, down slopes, and through sandy soils than it does in dry, level, or clay soils. Provided that affected plants are not allowed to grow rank, animals don't avoid eating plants growing in urine patches. In fact some animals prefer plants growing in urine patches, if they weren't burned by the urine. If urine scorches or kills plants, animals avoid eating them, and the botanical composition declines in quality. (see chapter 3)

Parasites

By constantly moving to fresh grazing, wild animals avoid serious damage from parasites. Domestic livestock are restricted to grazing a limited pasture area repeatedly during a season, however, and can be adversely affected by parasitism.

The most serious threat to livestock is from worm parasites of the digestive tract, although lungworms also can be a problem. Knowing the cycle of parasite biology and pattern of pasture contamination forms the basis of control. Animals become infected by grazing forage contaminated with infective parasite larvae. Inside the animal, roundworms only need about 4 weeks to develop to the stage of egg-laying adults. Infected animals shed parasite eggs in their dung, and within 2 weeks the eggs develop into infective larvae.

Reinfection of pasture occurs constantly during the grazing season. Conditions that favor pasture plant growth also favor development and survival of parasites.

Larvae can survive up to 10 months on pasture (eggs of Nematodirus roundworm of young cattle survive up to 24 months), so even the longest recovery period between grazings isn't enough to break the cycle.

Larvae that survive winter conditions immediately reinfect livestock when grazing begins in spring. Young stock are most susceptible to infection and the worst effects of parasitism.

Parasitism can cause large economic losses to livestock producers. Even it there are no obvious symptoms (subclinical), parasitism can depress an animal's dry matter intake, and its immune system. Because of this the animal grows more slowly, produces less, and becomes more susceptible to bacterial and viral infections.

Management
Parasite levels in pasture can be reduced and economic loss prevented by deworming at strategic times in the parasite life cycle and planning the grazing sequence so that treated animals can be moved to clean paddocks. Ask your veterinarian about dung exams to determine the level of parasite contamination of your pasture. If a problem exists, learn about the correct dewormer to use and when to use it in your situation. Be careful to avoid residue in milk or meat. Be aware that some dewormers kill beneficial insects and soil organisms.

The main objective in deworming livestock is to achieve pasture that's relatively free of parasites. Usually deworming should occur at 4 and 8 weeks after animals begin grazing a contaminated pasture.

Dewormed animals should be moved to land that's relatively free of parasites, if possible. Hot dry weather kills larvae, so land that's grazed by another animal species or machine-harvested in spring, is safe for treated animals to graze. Land that's seeded down after a tilled crop is free of parasites. Harrowing during hot weather reduces parasite levels.

Parasites are specific to animal species they infect, so alternate stocking by different species in different years can reduce parasites to low levels in a pasture. Alternately grazing and machine harvesting an area in different years gives similar results. Mixed species grazing (e.g. cattle with sheep) also reduces contamination levels.

In the north, where winter freezing/thawing conditions may reduce parasite egg and larvae numbers (except *Haemonchus*) on pasture, special grazing management can keep parasites at low levels in young stock. If young animals are with adults (e.g. lambs with ewes) graze each paddock only once per season with the young stock. If young animals are alone (e.g. dairy calves) graze rotationally. In both cases, the pasture cannot have been grazed by adults of the same species before in the current year. Adults shed eggs that quickly develop into larvae that infect young animals if they graze areas previously grazed by adults in the same year. After young animals have grazed a pasture once, during the rest of the season graze with other species or adults of the same species, or machine harvest the forage.

SEED DISPERSAL

Seeds attach to hooves, hides, and wool, spreading around a pasture as animals graze. Seeds are carried in digestive tracts of livestock, then dropped in dung. The degree to which eaten seeds are digested depends on both animal and plant species. Some animal digestive systems break down seeds more than others. Some seeds have harder seed coats that resist digestion more than others. About 10% of the seed that's eaten passes unharmed through animals.

Large numbers of seeds can be spread around by animals in their dung (10 live white clover seeds per gram of dung dry matter have been found). If these are seeds of desirable plants it benefits the pasture, but if they are undesirable they will only contribute to making the pasture more weedy. For this reason, it is always best to feed only forage that is relatively free of weeds. Weed seeds in hay that is fed during winter may pass through unharmed in dung, which is then spread on the land, creating weed-control problems. If possible, don't allow your livestock to graze a weedy pasture when the weeds have set seed, before grazing a clean pasture. Of course, if you manage your pasture well, weeds never flower!

SUCCESSION

Plant and animal (including soil organisms) succession moves naturally toward greater complexity and stability. In northeastern and northcentral USA and similar areas, energy resources (sunlight, temperature, precipitation) are high enough to support complex forest vegetation. Once forest is cleared, surplus resources are available. If they're not all used for pasture plant production, the extra resources will be used to grow woody plants and weeds.

Maintaining pastureland under these conditions requires periodic effort or work to hold back the successional movement to climax forest. The larger the energy resources of a region, the more frequently work must be done, or the more severe it has to be if done infrequently to bring the plant and animal community back to the young growth stage of a pasture. You can accomplish this either through grazing management with frequent moves of livestock at high stocking density and high utilization of pasture, or infrequent use of bulldozers, chain saws, and herbicides. The former method puts money in your pocket; the latter takes money from you.

REFERENCES

Gooding, A. 1980. Don't let your pastures waste money. Angus Journal. June-July. p. 348-350.

Hoke, D. 1993. *The Worm Turns.* University of Vermont Pasture Management Outreach Program. Mimeo. 4 p.

Johnstone-Wallace, D.B. and K. Kennedy. 1944. Grazing management practices and their relationship to the behaviour and grazing habits of cattle. J. Agric. Sci. 34:190-197.

Harris, W. 1978. Defoliation as a determinant of the growth, persistence and composition of pasture. p. 67-85. In Wilson, J.R. (ed.) *Plant Relations in Pastures.* CSIRO, Melbourne, Australia.

Myers, G.H. 1995. Deworm pasture. *Stockman Grass Farmer.* 52(4):23.

Nation, A. 1994. Dairy and sheep together make a good match. *Stockman Grass Farmer.* 51(9):7.

Savory, A. 1988. *Holistic Resource Management.* Island Press, Washington, D.C. 564 p.

Semple, A.T. 1970. *Grassland Improvement.* Leonard Hill Books, London. 400 p.

Smetham, M.L. 1973. Grazing management. p. 179-228. In Langer, R.H.M. (ed.) *Pastures and Pasture Plants.* A.H. & A.W. Reed, Wellington, New Zealand.

Smith, B., P. Leung, and G. Love. 1986. *Intensive Grazing Management.* The Graziers Hui, Kamuela, Hawaii. 350 p.

Vallentine, J.F. 1990. *Grazing Management.* Academic Press, NY. 533 p.

Voisin, A. 1959. *Grass Productivity.* Philosophical Library Inc., NY.

Watkins, B.R. and R.J. Clements. 1978. The effects of grazing animals on pastures. p. 273-289. *In* Wilson, J.R. (ed.) *Plant Relations in Pastures.* CSIRO, East Melbourne, Australia.

5
Voisin Management Intensive Grazing

Sitting silently, doing nothing,
Spring comes,
And the grass grows by itself.
                                        Zenrin

I don't think anyone would fill a bunk or trench silo with corn or alfalfa ensilage, and then turn livestock into it, allowing them to eat what and when they want to. Of course no one would do that intentionally, because everyone knows what would result: a mess! Instead the silage is carefully rationed out according to the needs of livestock, while keeping waste of the feed to a minimum.

But everyone who has grazed animals, including me, has turned them out into large pasture areas, allowing them to pick and choose, eating what and when they want to. The same kind of mess results. The animals tend to eat plants and parts of plants that they like best, leaving the rest to mature, set seed, and multiply. The most desirable plants, such as clovers, are grazed off every time they grow to grazing height. They never have enough time with enough leaf surface for photosynthesis to meet the needs of the plant. As a result, these plants soon wear out and die. With this kind of management, any pasture becomes weedy and unproductive.

In the United States this hasn't been recognized as a management problem; instead, the pastures have been blamed. Not many people would try to grow corn, soybeans, wheat, or alfalfa with the same level of attention to cropping management and soil fertility that they use on their pastures. If they did, yields of those crops wouldn't amount to much of anything. But permanent pastures in the USA traditionally have been managed according to the Back-40 Syndrome: don't put anything in, don't expect anything out.

The Back-40 is the run-down permanent pasture where animals are placed in spring and survivors are collected in the fall. If most survive, the farmer congratulates him/herself for doing a good job of management. Some animals even gain weight if the pasture was well understocked. When animals requiring a high level of nutrition, such as milking cows or growing lambs, are grazed with similar management, they aren't very productive.

It's no wonder that farmers in the United States stopped pasturing their animals and went to year-round confinement feeding: the yield of livestock products from pasture was too low to be profitable. But the ironic part is that it never was the pasture's fault. It always resulted from poor management of a forage crop in a pasture situation.

If it weren't for New Zealand's highly productive and profitable agriculture, which depends almost entirely on permanent pastures, American farmers probably would have gone on forever, blissfully ignorant of better grazing management techniques and spending much more than they need to in producing livestock. But New Zealanders profitably produce lamb and dairy products, ship them halfway around the world, and underprice American farmers. They are doing something right. But what?

New Zealanders applied, adapted, and refined the ideas of Andre Voisin to their conditions with great success.

Combining his scientific training and experience with diligent observation of livestock in the field, Voisin discovered that *time* was crucial in grazing management. Like many people who have made important discoveries, Voisin didn't receive the credit and recognition that he deserved, especially in the United States. Without the concept of time we still wouldn't know how to properly manage grazing livestock.

When Voisin defined the concept of time in grazing management, some important things were realized, especially about overgrazing and stocking density. It became possible to minimize overgrazing and undergrazing. He showed that overgrazing is unrelated to the number of animals present in a pasture, but is highly related to the time period (how long and when) during which plants are exposed to the animals. If animals remain in any one area for too long and graze regrowth, or if they return to an area before previously grazed plants have recovered, they overgraze plants (more of this discussion under Plant Roots, chapter 2). For example, 200 cows grazing a 1-acre paddock (that has adequately recovered from a previous grazing) for 12 hours don't overgraze, but one cow grazing the same 1-acre paddock for 7 days or more does overgraze.

Voisin's method of planned, management intensive grazing interferes as little as possible with the pasture environment, while gently guiding it to benefit the farmer, and protecting it from damage by grazing animals. It is a simple method that in essence just gives pasture plants a chance to photosynthesize and replenish energy reserves after each grazing. Using this method, you control what and when livestock eat, by dividing pastures into small areas (paddocks) and rotating animals through them. You can then ration out pasture forage according to the needs of livestock, allow the plants to recover from grazing according to their needs, and keep forage waste to a minimum.

Key parts of Voisin management intensive grazing concern rest or recovery periods between grazings, and the length of time that animals are in a paddock. Let's look at those parts and related aspects.

TERMINOLOGY

First I want to introduce some useful terms that can help

in discussing and understanding what's going on in a pasture.

Pasture mass: total weight of forage per acre, measured to ground level, and expressed as pounds of dry matter per acre (lb DM/acre). When measured before grazing it's called "pregrazing pasture mass"; measured after grazing it's called "postgrazing mass, residue, or residual."

Cover: average pasture mass over a pasture.

Forage allowance: total dry weight of forage provided, measured to ground level, and expressed per animal or per unit of animal liveweight (e.g. lb DM/cow or lb DM/100 lb liveweight).

Net forage: the amount of forage that actually can be harvested from a pasture (lb DM/acre), either as forage consumed by grazing animals or removed by cutting. Rate of net forage production indicates how much of that harvestable amount is produced per day (lb DM/acre/day).

Stocking rate: number of livestock carried or supported per acre during a season or part of a season (e.g. 1.4 cows/acre). It is directly related to the amount of feed grown per acre. Stocking rate is important in balancing the feed requirements of livestock with the amount of pasture forage grown on a farm (see chapter 8).

Stocking density: the concentrated number of animals grazing a paddock at a given moment, expressed as number of animals per acre per time period (e.g. 200 cows/acre/12 hours). Stocking density greatly influences pasture plant growth and forage use. Through its effect on herd or flock behavior and level of feeding, it also affects animals' feed conversion efficiency. If the number and/or size of animals remain the same, decreasing paddock size increases stocking density. Increasing numbers and/or size of animals while paddock size remains constant, increases stocking density. And vice versa.

PASTURE SWARD DYNAMICS

To manage pastures well, the biological and ecological bases for management must be understood. Then grazing management can be flexible, based on observation of plants, soil, and animals, and understanding what's going on, rather than rigidly following a set schedule of calendar dates that may have very little or nothing to do with pasture plant growth.

Basically, it all boils down to the fact that pasture plants must be able to regrow after they have been grazed. The regrowth is powered by energy either from photosynthesis occurring in remaining leaf surface, or from energy reserves in roots, stolons, and stem bases if little or no leaf surface remains. Except for alfalfa (which regrows from energy reserves) and after drought, pasture grasses and legumes regrow mainly from photosynthesis occurring in remaining leaf surface. Regrowth from energy reserves is slower because it takes time to form enough leaf surface so that photosynthesis can function at a high rate again. Photosynthesis supplies energy for continued regrowth of the plant and storage of more energy reserves.

Low stocking density and poor forage utilization (lax grazing) result in sparse swards having few or no lower leaves on plants because of shading; when grazed closely, such plants regrow slowly because little or no leaf surface remains. So if plants are cut or grazed before enough reserves are stored, and there are few or no lower leaves remaining, regrowth will be retarded or won't occur at all.

Plant Tissue Flow

Underlying the above simple overview above is a complex interrelationship of plants, temperature, light, soil, organisms, nutrients, water, and livestock that makes pasture a continually changing (dynamic) community. In a pasture there is a continuous flow of new plant tissue forming and old tissue disappearing through the processes of aging, death and decay, or consumption by grazing animals (Figure 5-1). This turnover of tissue can occur very fast. For example, in perennial ryegrass a new leaf appears on each tiller about every 11 days. Ryegrass plants maintain only three live leaves per tiller, so each leaf lives an average of just 33 days. If leaves aren't eaten within their short lifespan, they are lost to the grazing animal.

Climate, soil fertility, and plant species mainly determine the rate of new herbage formation in a pasture sward. Grazing management can influence this rate by maintaining as green and leafy a sward as possible, so photosynthesis can occur all the time, including right after grazing when the plant residue is very short. Sparse, stemmy, yellow postgrazing plant residue with little green leaf surface area remaining, such as occurs when plants have been allowed to grow too tall, takes longer to reach a maximum rate of new herbage formation than does a dense, leafy, green residual.

Plant Regrowth Curve
The regrowth curve (Figure 5-1) of plants is S-shaped and has three stages:
1) early period of slow growth,
2) middle period of rapid growth, and
3) final period of slow growth.

In the first growth of spring or after being grazed or cut off anytime in the season, plants have limited leaf surface area and can only grow slowly. As leaf surface develops and light interception increases in full sunlight, pasture plant growth increases quickly, then gradually slows as pasture mass and shading increase. The rate of new plant growth reaches a maximum at a pasture mass of about 800 to 1100 lb DM/acre in dense, leafy swards grazed by sheep.

Pasture grazed by cattle is less dense and more clumpy, with more erect tillers and leaves than sheep or goat pasture. Because less light is intercepted, cattle pastures need more leaf surface area and a pasture mass of 2200 to 2700 lb DM/acre to reach maximum growth rate.

Net Forage Production

After grazing, the rate of net forage production increases rapidly as plant growth and pasture mass increase. When high levels of pasture mass are reached, net forage production decreases as pasture mass continues to increase. This decrease reflects the smaller proportion of new plant growth being harvested, as more forage is lost through death and decay from shading of lower plant parts and low-growing plants such as white clover.

Shaded plant parts continue to respire, but without light they can't photosynthesize, so until they die they draw energy from parts exposed to the sun. This is a direct loss to your pasture's productivity, and is the reason why swards must be kept below 6 to 8 inches tall!

Sooner or later as pasture mass continues to accumulate under poor or very lax grazing management, net forage production becomes zero (i.e. all of the forage rots and animals won't eat it), and pasture mass can't increase any more because of shaded conditions within the sward. There is a range of pasture mass over which net forage production remains high. The rate of net forage production drops markedly at extremely high (3000-4000 lb DM/acre) and low (400-600 lb DM/acre) levels of pasture mass (Figure 5-1).

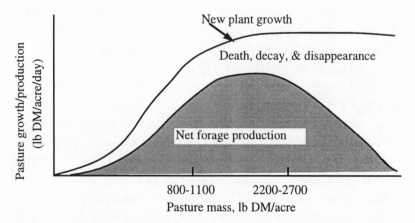

Figure 5-1. Influence of pasture mass on rates of new plant growth, net forage production, and forage loss through death, decay, and disappearance (Adapted from Korte, Chu, and Field 1987).

RECOVERY PERIODS

Besides variation of the plant regrowth curve, plant growth rate also differs within the season. One of the main rules of Voisin management intensive grazing is that recovery periods between grazings must vary according to changes in plant growth rates, which reflect changes in growing conditions. In general, this means that recovery periods must increase as plant growth rate slows as the season progresses. For example, in the northeastern and northcentral United States plant growth rate in May through June is about twice as fast as it is in August through September, and July is a transition time between the two. This means that recovery periods between grazings must be about twice as long in August-September as they are in May-June. Of course, plant growth rates vary within regions and prevailing climatic conditions in any season. (Figure 5-2)

Productivity of plants and the amount of forage available to animals entering a paddock, equals the daily amount of plant regrowth per acre (pasture mass) that accumulated since the last time the paddock was grazed.

Looking closely at Figure 5-2, you can see that if the recovery period is cut to half of the optimum period, forage accumulation is reduced about two-thirds. If the recovery period is shortened even more, forage accumulation may drop to only 10 percent of the pasture mass accumulated during the optimum recovery period. This short recovery period corresponds to what happens when pastures are continuously grazed: desirable plants are grazed off every time they grow tall enough to be grasped by the animals' mouths, or about every 7 days. If recovery periods are longer than the optimum, pasture mass increases, but the increase is due mainly to more fiber, which lowers the feeding value of the forage.

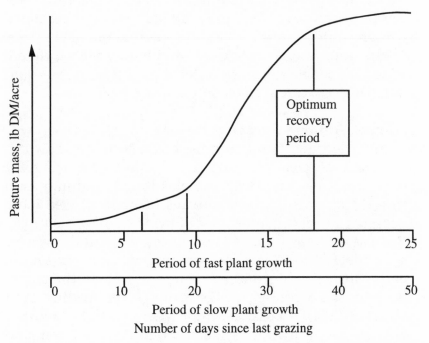

Figure 5-2. Relationship of recovery period to pasture mass accumulation during periods of rapid (e.g. May-June) and slow (e.g. August-September) plant growth (Adapted from Voisin 1959)

Basic recovery period guidelines are helpful for planning and beginning to use management intensive grazing. As you gain experience, you can make adjustments to better suit your local conditions and pasture plant communities. For example, in the Champlain Valley of Vermont, we have found that these recovery periods work well for Kentucky bluegrass-orchardgrass-timothy-quackgrass-white clover pasture:

• 12 to 15 days in late April to early May
• 18 days by May 31
• 24 days by July 1
• 30 days by August 1
• 36 days by September 1
• 42 days by October 1

Recovery periods needed in other climatic zones or for other pasture swards (e.g. grass-alfalfa) can be determined by taking into account what is known about the plants' carbohydrate reserve cycles, experimenting with different periods, and observing the effects of different pre- and postgrazing pasture masses on plant regrowth rates.

Spring-to-Autumn Recovery Periods

An important question that must be answered is: how can the length of recovery periods be adjusted during the season until they are about twice as long in autumn as in spring? There are 3 practical ways of doing it (see more discussion in this chapter under Spring Management):

• Because too much forage is produced early in the season, set aside 1/2 to 2/3 of the pasture area from grazing in early spring and conserve surplus forage (machine-harvest and store as hay or haylage for feeding later when needed). If possible, cut it when it's 10 to 12 inches tall, so low-growing legumes aren't shaded out and there's little stubble later to bother grazing animals. The forage you harvest at this stage will be very high quality. Top-grazing the set-aside area once in early spring will shift the surplus to later in the season when the weather is better for making hay.

After plants in the machine-harvested area regrow to your required pregrazing mass, incorporate as much of the area needed to increase recovery periods to what is adequate for that time of season (e.g. late spring-early summer). Divide this amount into paddocks and include them in the rotation; this increases the area available for grazing and automatically lengthens recovery periods.

Plants in the rest of the area can be allowed to grow to 10 to 12 inches tall and conserved again, before beginning to be grazed for the first time in late summer. This means that only about one-half or less of the total pasture area is grazed in May, June, and part of July!

Because using management intensive grazing at least doubles or triples plant productivity, you'll have to conserve forage from areas where surplus forage probably has not been harvested before. If your pastures are mainly on rough land, set aside the most level areas where you can use machinery to harvest the surplus. You must be prepared for the increase in forage production that will occur, otherwise the pasture won't be grazed properly and its full potential won't be realized.

If possible, conserve surplus forage from different areas each year to avoid thinning the sward and decreasing legume content due to shading by the high pasture mass that accumulates before machine harvest. Be sure to apply manure to these areas to replace nutrients removed in the conserved forage.

For example, on Mike and Tammy Hanson's farm in Vermont, we fed 60 lactating Holsteins and 15 dry cows and heifers on 15 acres of pasture from April 29 to about June 15. They have a total of 50 acres of pasture. In late May, Mike conserved surplus forage as haylage from the remaining 35 acres of pasture. In June we brought about 15 acres of that machine-harvested land into the grazing rotation. In July Mike conserved a second crop of surplus forage as hay from the other 20 acres. In September we included all 50 acres in the rotation. Cows grazed until mid-October; heifers and dry cows grazed until about November 1.

• Graze more animals on the total pasture area during the first half of the season than you can feed during the rest of the season. At mid-season decrease as much as necessary the number of animals carried, and this will lengthen recovery periods to about what is needed. This really is the only way to graze a pasture well if it is all rough land that can't be harvested with machinery anywhere. This means that about half of the animals will either have to be fed elsewhere after July 15, or will have to be sold.

• Another way of keeping pasture forage under control during times of rapid, excessive growth, especially on land where machine harvesting isn't possible, is grazing less close (top-graze) than usual, leaving more residual. This means that you either rotate animals through paddocks faster than usual (e.g. move animals several times a day), or allow animals to graze several or all paddocks at once. The patchy grazing that will occur isn't a serious problem if only allowed occasionally, and can be minimized by grazing two kinds of animals (e.g. sheep and cattle) on the same land, either simultaneously or one behind the other. When plant growth rate decreases, slow the rotation and return to grazing one paddock at a time.

Adjusting Recovery Periods

Recovery periods are, of course, based on observation of plant regrowth and pasture mass. Under conditions that stress plants, such as drought or cold, longer recovery periods are needed. If conditions are more favorable (e.g. warm, moist) than usual for plant growth, less recovery time may be needed.

For example, depending on the animals and their production levels and nutritional needs, in Vermont we allow swards of Kentucky bluegrass, orchardgrass, timothy, quackgrass, and white clover to accumulate pregrazing pasture masses of 2100 to 2200 lb DM/acre (3-4 inches tall) for sheep, and 2100 to 2400 lb DM/acre (4-6 inches tall) for dairy cows in each rotation, before animals are turned in. We remove producing animals from paddocks when swards are grazed down to about 1400 lb DM/acre (2 inches tall). Following groups of dry ewes or dry cows and heifers graze down to 1000 to 1200 lb DM/acre (1-1.5 inches tall). These pre- and postgrazing masses are target averages; parts of the sward in a paddock will be above or below these levels.

More dense grass-legume mixtures, such as perennial ryegrass-white clover, can accumulate more pregrazing mass (e.g. 2700-3000 lb DM/acre) without damaging the sward and decreasing forage consumption. This is because they're still short and leafy enough for easy grazing and for sunlight to reach deeply even at high pregrazing mass.

If the length of time needed to graze a paddock to the required residual pasture mass decreases, the recovery periods for the next paddocks also will be shortened if animals are moved sooner than planned. If recovery periods begin to shorten by even 12 hours in any rotation, it is a warning signal that plant growth has slowed for some reason.

For example, around June 20 you notice that recovery periods, which had been 22 days, begin to shorten by 12 hours in each paddock. Carefully check the pasture mass in paddocks that the animals will go into next. If the plants haven't regrown enough in the next paddock, you must slow the movement of animals through the rotation to allow plants more recovery time. You can accomplish this by:

• increasing the pasture area and number of paddocks available for grazing,

• removing all animals from the pasture and feeding them elsewhere, or

• feeding the animals hay or other forage in a paddock or paddocks until enough pasture mass accumulates (2100 to 2400 lb DM/acre) and recovery periods are adequate again.

(Anytime that you feed hay, greenchop, or silage to animals, including during winter, try to do it on a different area of your pasture every day. This improves pasture soil fertility from added manure and wasted forage, saves you the time and expense of cleaning up and spreading manure, and adds new forage seeds to the pasture. See chapter 9)

Another way to increase recovery periods early in the season is to graze areas that had been set aside for machine harvesting, even though the surplus forage hasn't been cut yet. The animals won't like having to graze tall, more mature plants, after being accustomed to grazing short immature plants, and they will waste a lot of forage if you let them. To prevent waste, give them narrow strips of forage about 10 feet wide, or only enough for 12 to 24 hours. Hold the animals in the strips until they eat most of the forage. Then if possible mow each strip so that they can eat the remaining forage before giving them another strip. It's hard for ruminants to bite off stemmy plant parts, so you help them by mowing the plants in this situation.

Animals needing high levels of nutrition (e.g. milking cows, sheep, or goats, growing lambs) can't be forced to graze this more mature forage completely without reducing their production, however. So don't force animals with high nutritional needs to graze it as closely. But mow the area that is used for grazing while increasing recovery periods, at least once, either in every strip or as soon as you remove the livestock from the area. Mowing cuts the stems that weren't eaten and evens the plant growth for grazing in the next rotation.

If you have to increase recovery periods later in a season, your animals can graze hayland aftermath, corn, sudangrass, millet, or other crops seeded for this purpose. You also can feed green chop or feed some of the surplus forage harvested earlier from the pasture.

If plants in the next paddocks to be grazed don't recover enough, less forage is available and, consequently, the movement of animals accelerates through the rotation at a time when it should be slowing down. Voisin referred to this faster movement of animals as "untoward acceleration". Quite suddenly plants become exhausted, stop growing, and there is no more forage to graze.

--∞--

Watch plant regrowth and pasture mass: they are your best indicators of what conditions are like for the plants. Movement of animals among paddocks must always be based on observation of plant regrowth and pasture mass. By meeting this need for adequate recovery time between grazings, pastures can be grazed for most of the year, especially with autumn forage stockpiling (see chapter 9).

PERIODS OF STAY AND OCCUPATION
The length of time that each group of animals is in a paddock per rotation is called the period of stay. The total time that all groups of animals occupy a paddock in any one rotation is called the period of occupation. If only one group of animals grazes a paddock system, the period of stay equals the period of occupation. If two groups graze, their total periods of stay equal the occupation period.

When plants are grazed off, they can regrow tall enough to be grazed again in the same rotation if the occupation period is too long. So periods of stay or occupation must be short enough to prevent grazing of regrowth. In the northeastern and north central United States, plant regrowth may be tall enough to be grazed again after about 6 days in May-June and 12 days in August-September. Although most plants take 12 days in August-September to regrow to grazing height, some continue to regrow tall enough to be grazed after only 6 days. For that reason, periods of occupation should never be longer than 6 days to prevent grazing of regrowth in the same rotation, and really should be 2 days or less for best results.

Periods of stay for any one group of animals shouldn't be longer than 3 days, giving total occupation periods of 6 days for two groups of animals. This is because the longer animals are in a paddock, the less palatable the remaining forage becomes, and the more time and energy they spend searching for desirable feed. Periods of stay of 2 days or less if animals are grazed as one group, and 1 day or less for each of two groups, giving total occupation periods of 2 days or less, are better than longer periods of stay and occupation.

In practice then, the shorter the periods of stay and occupation, the better the conditions are for optimum plant and animal production. Milking and growing animals shouldn't be in a paddock for longer than 2 days per rotation anytime in the season, to keep them on a consistently high level of nutrition. Milking cows, sheep, and goats produce the most milk if they are given a fresh paddock after every milking. Not only is the forage of higher quality and grazed more uniformly than with less frequent moves to fresh paddocks, but milking animals let their milk down better, anticipating that they're going to a fresh paddock as soon as they're done milking. Growing animals, such as lambs and stockers, also gain weight most rapidly if given a fresh paddock every 12 or 24 hours.

Paddocks must be small enough so that all forage in each paddock is grazed completely and uniformly within each occupation period. Occupation periods may need to change because plant growing conditions vary during the season, and the amount of forage available changes.

Lengthening occupation periods, and pregrazing pasture masses consistently above 2700 to 3100 lb DM/acre indicate that surplus forage is available. When that happens, some paddocks should be removed from the rotation and machine-harvested, so that the pasture continues to be well grazed to maintain forage quality, sward density, and plant regrowth potential. An alternative is to move animals through the rotation quicker, grazing less closely until the rate of plant regrowth slows.

Shortening occupation periods and pregrazing pasture masses below about 2000 to 2200 lb DM/acre indicate that plant growth rate has slowed and not enough forage is accumulating. In this case more paddocks and pasture area are needed.

For example, if animals don't eat enough to keep up with the rapidly growing plants in May and June, remove paddocks from the rotation and cut the forage for hay or silage. Suppose that in the first rotation of the season your animals occupy paddocks for 2 days, eating all of the forage available in each paddock within the 2 days. After they have grazed six or seven paddocks, move them to the first paddock that was grazed and start the second rotation. Leave the rest of the pasture area for machine harvesting.

(Return animals to the first paddock only if the plants in it have regrown to at least 4 inches tall and have a fully developed green color. Never allow animals to graze forage that's so young that it is still yellow. Grazing young yellow plants has a similar effect on animals that eating green apples has on people! Grazing plants that are too immature also may result in the plants growing poorly during the rest of the season.)

Recovery periods and occupation periods should work together so that the forage is at the right pasture mass or height when animals are turned into a paddock. When animals enter a paddock, plants should be 2100 to 2400 lb DM/acre (4-6 inches tall) for cattle or horses, and 2100 to 2400 lb DM/acre (3-4 inches tall) for sheep, goats, pigs, or poultry.

When animals leave a paddock, ideally all or most plants should have been grazed down to 1000 to 1200 lb DM/acre (1-2 inches tall). Notice the word "ideally" in the last sentence. We are working with animals that don't seem to have the slightest idea that we would like them to graze closely and uniformly. Their natural inclination is to eat the most palatable forage and leave the rest, especially forage near their own dung.

Depending on plant species and local environment, postgrazing plant heights and pasture masses may need to be greater. But usually the taller the remaining plants are when animals are removed from a paddock, the more selective and uneven the grazing has been. In the early stages of pasture improvement, it may be very difficult to have animals graze closer than 2 inches from the soil surface, because of stubble and matted plant parts from previous years of poor forage utilization.

It always will be difficult to get animals to graze closely in areas that have been machine-harvested, probably because the dry fibrous stubble remaining after mowing picks the animals' mouths. One way to avoid leaving stubble is to machine harvest surplus forage twice, or every time the sward reaches about 10 inches tall, rather than once when the sward is 18 or more inches tall. Two or more immature harvests result in leafier, higher quality hay or silage, and less loss of low-growing plants from shading. Of course, if it's difficult to dry early cut forage in your area, you may be limited to one harvest taken later. One way to delay surplus forage harvests until better drying conditions exist, is to graze surplus areas in the first rotation, then set them aside for later cutting (s).

Animals with high nutritional needs (e.g. milking cows, sheep, or goats) probably shouldn't graze forage down to 1 inch (900 to 1100 lb DM/acre postgrazing pasture mass) from the soil surface, unless the sward is very dense and leafy because their production may be lowered. This is why it is best to follow producing animals with animals having lower nutritional needs (e.g. dry cows, heifers, dry ewes) to graze paddocks down closely (discussed below).

If for some reason you can't graze with two groups of animals, don't force producing animals to graze down to less than 2 inches unless you are willing to lose some production. If your pasture can be mowed, producing animals can be allowed to graze less closely, and the pasture can be mowed after each grazing or periodically during the season. This is a compromise though, because, while your animal production will remain high, the pasture won't reach its full potential that ultimately could have resulted in even higher levels of animal production. Running machinery to mow or do anything else also takes time, costs money, and burns fuel.

Just as you must protect the pasture from being overgrazed by using short occupation periods, you should make certain that it isn't undergrazed as well. Any plants that aren't grazed in one rotation, probably won't be grazed again that season unless they are clipped. It saves you time and money, and net forage production increases if the animals graze as uniformly as possible down to 1 to 2 inches from the soil surface.

STOCKING RATE AND DENSITY

High stocking density, coupled with quick, close grazing of a paddock, and removal of animals until plants recover simulates what happened under natural conditions during millions of years. Then large herds of grazing animals grouped closely because of predators would move through an area eating and trampling plants and disturbing the soil. Their passage broke up plant residues and returned them to the soil to be decomposed. Disturbing the soil allowed seedlings to establish and water to enter. Large amounts of manure left behind provided nutrients for plant regrowth. By the time the herd returned the plants had recovered and were ready to be grazed again. Pasture and range plants evolved under this kind of treatment and require it for best growth. This is the underlying reason why Voisin management intensive grazing has such beneficial effects on pasture: it fits with the natural way of things.

It follows that stocking rate and density greatly affect pasture forage use and net forage production. Increases in stocking rate and density always reduce the amount of pasture forage that is wasted, thereby increasing the efficiency of forage use and net forage production.

On the most intensively managed and efficient farms, for example, livestock eat 80 to 90 percent of forage available to them during the season. On less well-managed farms, half of the forage produced may not be eaten, mainly because of low stocking rate and density. Wasting pasture forage prevents getting high levels of plant and animal production per acre.

Increasing stocking rate and density results in pastures being grazed more intensively during most of the season. Benefits of intensive grazing include less dead and dying herbage in the pasture (i.e. net forage production goes up), better forage digestibility, and more white clover and tillering of grasses in the sward.

Of course, when you increase the stocking rate on your farm, more livestock graze per acre and per ton of pasture forage produced. Less forage will be wasted, but eventually you could increase to a stocking rate where the amount of forage eaten per animal decreases, and its production level consequently declines. It's relatively simple to estimate what the stocking rate should be, based on animal energy and dry matter intake needs and pasture forage production (see chapter 8). Pasture stocking rates generally range from 0.7 to 1.8 animal units per acre (1 animal unit = 1000 pounds liveweight).

Stocking density depends on how much forage is available to animals in a paddock, how much of it you want them to eat, and how long you want them in the paddock. Paddock size and stocking density should combine so that animals don't have to be moved more frequently than twice a day, or less frequently than every 6 days, as discussed above. Keep in mind that more intensive management produces the highest pasture and animal yields. Stocking densities generally range from about 25 to 200 animal units per acre per occupation period of a paddock, depending on management.

FORAGE ALLOWANCE

Livestock must be offered amounts of pasture forage that are two to three times more than what they will eat, depending on the animals' production level and nutritional needs. This is to ensure that they are able to eat as much as they possibly can. For example, if a milking cow needs to eat 35 lb DM/day, she must be offered a forage allowance of 70 to 105 lb DM/day. She will leave 40 to 60 percent of the amount offered uneaten as postgrazing pasture mass. So livestock that need to eat large quantities of forage usually must be allowed to leave relatively large amounts of postgrazing pasture mass (1200 to 1400 lb DM/acre). Animals with low nutritional and dry matter intake needs can be given a forage allowance that requires them to graze to a lower postgrazing pasture mass (1000 lb DM/acre); these can follow animals with high nutritional and dry matter intake requirements.

SWARD MEASUREMENTS

Besides understanding the biological and ecological reasons for what you do, you must be able to estimate the amount of forage present in a sward before and after grazing. It's helpful to discuss sward measurements before getting into other details of management intensive grazing.

Plant Height, Density, And Pasture Mass

Plant height and density combined are included in pasture mass. For a particular pasture botanical composition and time during the grazing season, pasture mass is useful for predicting plant and livestock performance.

Pasture mass interrelates with forage quality and palatability in affecting productivity of grazing livestock. High pregrazing pasture mass over 2700 to 3000 lb DM/acre (10-12 inches tall) occurs with low stocking density or infrequent grazing. These high levels of pasture mass can quickly result in decreased forage quality, patches of rank, low-quality uneaten forage, a layer of dead and decaying plant material at the base of the sward, more upright growing plants, patches with no live tillers, death of low-growing legumes, and a shift to undesirable plant species.

For maximum plant and animal production, pasture mass generally should fluctuate between 1000 to 1100 (postgrazing) and no more than 2700 (pregrazing) lb DM/acre, depending on the kind of livestock. Occasional close grazings down to 800 to 1000 lb DM/acre are needed to maintain high forage quality and desirable sward composition.

The main objective of management intensive grazing is to keep pasture plants within the steep part of their growth curve, so that the rate of new plant growth always remains high (Figure 5-1). If they're kept within the steep part of the curve, regrowth occurs rapidly after plants are grazed. If postgrazing pasture mass reaches less than 800 lb DM/acre, plants have little leaf surface and are then in the low part of the curve, and regrowth occurs slowly until adequate leaf surface develops. For this reason, pre- and postgrazing pasture masses of each paddock must be estimated, in deciding when to move animals in and out of paddocks.

Pasture mass usually is estimated visually (eye-balling) or by measuring plant height (sward surface height). Sward height measurements or estimates must be calibrated against actual measurements of the forage present in an area. Sward height affects plant growth and death through shading and, consequently, influences net forage production. Sward height also affects forage intake and performance of grazing livestock.

For example, maximum utilization of perennial ryegrass-white clover by grazing livestock occurs when swards are maintained at 1.5 to 2.5 inches tall for sheep, and 3 to 4 inches tall for cattle! For other cool-season grasses (e.g. Kentucky bluegrass, orchardgrass) combined with white clover, best pregrazing sward surface heights appear to be 3 to 4 inches tall for sheep, and 4 to 6 inches tall for cattle. Optimum postgrazing sward heights are 1 inch tall for low-growing grasses (e.g. perennial ryegrass, Kentucky bluegrass), and 2 inches for tall-growing grasses (e.g. orchardgrass, timothy, bromegrass).

Using sward height to guide grazing management isn't completely satisfactory, however, because it doesn't take into account variations in plant species and density. For uniform swards, such as nitrogen-fertilized perennial ryegrass growing under uniform conditions, sward height correlates closely with plant density and pasture mass. But sward height measurements may be misleading for estimating the pasture mass of complex swards that contain several grass species, depend on a legume for nitrogen, and grow under variable conditions, such as in most of the northern United States.

Since plant density is more difficult to estimate than plant height, visually estimating pasture mass requires experience and also some actual measurements of the forage to calibrate visual estimates. Visual density and height estimates may be checked against the actual total amount of forage present, by first estimating the pasture mass in pounds of dry matter per acre and measuring plant height within several 1- x 3-foot areas. Then cut the forage within the areas at the soil surface and place the forage from each area in a paper bag that you have noted your estimates on. Dry the samples in an oven at about 160 degrees Fahrenheit for 1 day, and weigh (in pounds) the dry forage. Convert the dry weights to pounds of dry matter per acre by multiplying each dry sample weight by 14520. Compare your estimates to actual amounts of forage present.

Easier and more accurate ways of estimating pasture mass, especially in complex swards, would be to take about 30 readings throughout a paddock with a pasture bulk height plate or with an electronic capacitance meter, both of which take plant density and height into account.

We have been using a pasture plate developed by Ed Rayburn (see references) that is very inexpensive and simple to make and use for estimating pregrazing pasture mass. (Because of its heavy weight, this plate doesn't work well for estimating postgrazing mass.) Just buy an 18-inch square piece of 1/4-inch thick acrylic plastic and a yard stick at a hardware store. Cut a 1.5-inch diameter hole in the center of the plastic square. (Cutting the hole is the most difficult part, because the plastic tends to melt back together. Try to have someone cut the hole at the hardware store.) Attach a 1/8- x 2.5-inch bolt through a hole in the lower end of the yardstick, to pick up and carry the plastic plate easily as you walk around your paddocks. Now insert the yardstick through the hole in the plastic plate, and you're all set!

Start taking readings near one corner of a paddock in an area that will be grazed. Always estimate pasture mass in representative areas that animals will graze, not in rejected forage areas around dung. Gently lower the flat side of the plate with the upright yardstick onto the sward. Push the yardstick down until it just touches the soil, and allow the plate to settle by its own weight. Then bend over and look across the top surface of the plate to read the ruler measurement of forage bulk height. Record the reading to the nearest 1/4 inch on a paper or on a calculator that sums and tells you how many entries you've made. When you've finished taking about 30 readings by zig-zagging across the paddock, calculate the average forage bulk height. If your sward is composed of white clover and grasses such as Kentucky bluegrass, orchardgrass, timothy, bromegrass, and quackgrass, you can multiply the average bulk height by 432 to estimate pounds of dry matter per acre.

For best results the plate should be calibrated so that you know how many pounds of forage dry matter per acre the bulk height readings correspond to for your swards. To do this, make a wire frame that just fits over the plate. At several sample locations in your paddocks place the wire frame over the plate after you have taken and recorded a bulk height reading, and remove the plate. Then arrange the plants so that the wire frame lays flat on the soil surface. Plants rooted within the frame should be in the frame, and plants rooted outside should be outside of the frame. Then cut all of the plants within the frame at ground level with a scissors or battery powered lawn edger. Place the plant material in a paper bag, and note paddock number, date, and bulk height reading on the bag. Dry the samples in an oven at about 160 degrees Fahrenheit for at least 24 hours. Then weigh the samples and multiply the dry weights by 19360 if you weighed in pounds, or 42.6 if you weighed in grams, to get pounds of dry matter per acre.

Now graph the values, with inches of bulk height on the X-axis and pounds of dry matter per acre on the Y-axis. Future bulk height readings can be compared to the graph, to estimate dry matter per acre.

Another way of estimating pasture mass is with a computerized capacitance meter that uses changes in electrical capacitance resulting from different plant surface areas, to measure pasture mass (e.g. Pasture Probe, Pasture Gauge). With a capacitance meter, you can accurately determine when paddocks are ready to be grazed, when animals should be removed from paddocks, and daily and seasonal forage yield of each paddock and the entire pasture.

DIVIDING THE ANIMALS

When your pasture is well developed and leafy and swards are kept short (less than 6-8 inches tall for cattle; less than 4 inches tall for sheep), forage quality is high throughout the range of pasture mass, so producing animals can graze down to required low residual mass without losing production. Then the simplest way to graze is to include all animals in one group. Having the animals all in one group also gives the highest stocking density.

Although only one source of water is needed when animals graze as one group, ideally drinking water should be provided in all paddocks, so that the animals and their manure remain in paddocks being grazed. Also, if drinking water is always readily available in paddocks, animals don't waste energy walking to and from the source of water.

During the early stages of pasture development, it's usually best to divide animals into groups according to their production levels and nutritional needs at different physiological states. This allows you to closely match pasture feeding value with animal needs. For example, a dairy herd can be divided into two groups of milking animals and dry animals with young stock. A sheep flock or beef herd can be divided into two groups of weaned animals and dry mothers.

The groups of animals can be handled in two ways:
- Each group can be grazed in separate cells. (A cell is an area of land planned for grazing management as one unit, subdivided into paddocks to ensure adequate timing of grazing and recovery periods.) Animals with the highest nutritional requirements (e.g. milking cows, sheep, and goats, growing lambs or stocker cattle) are grazed on the best pasture available. Other animals (e.g. dry cows and heifers) graze a separate cell on lower-quality pasture that's being improved.

An advantage of this method is that only one source of

water is needed in each cell, although providing water in all paddocks is preferable. Another advantage is that parasites aren't easily passed from older animals to young ones. For producing or growing animals, paddocks should be small enough so that all forage is eaten in each paddock in 2 days or less (12-24 hours is best). Occupation periods for the animals at lower levels of nutrition can be 6 days or less, but they and the pastures will do better if 2-day or less periods of occupation are used also.

• Both groups graze within the same cell (leader-follower grazing). Animals having the highest nutritional needs are turned into a paddock first so they can eat the best forage quickly. They should not be left in a paddock for more than 2 days, because after that time they have to work too hard to meet their nutritional needs (12- to 24-hour periods are best).

After the first group is removed from a paddock, the second group follows to clean up the remaining forage, which has a lower feeding value than the forage that was grazed first. Paddocks must be small enough so that the combined periods of stay of the two groups are less than 6 days (2 days or less is best).

When two groups of animals graze within the same cell, all paddocks must have a source of drinking water, because at least one of the groups must stay in its paddock to keep the groups separate.

NUMBER OF PADDOCKS

When planning your grazing management, you need to estimate how many paddocks will be required to ensure adequate recovery periods between grazings. The number of paddocks needed depends on recovery periods, stay or occupation periods of animals in each paddock in each rotation, and number of groups grazing (Table 5-1).

Since shorter occupation periods favor higher plant and animal yields, the more paddocks you have up to a

certain point, the more productive your pasture will tend
to be. In deciding how many paddocks to build, you
should consider topography and soil fertility of the
pastureland, pasture sward botanical composition and its
potential yielding ability, maximum recovery periods
likely to be needed in your area, your livestock, fencing
costs, and your financial situation.

Table 5-1. An example of the number of paddocks needed
for a 36-day recovery period between grazings.

| Period of stay for 1 group, days | Total number of paddocks needed for | |
	1 group	2 groups
1/2	73	74
1	37	38
2	19	20
3	13	14

First, estimate the recovery period likely to be needed
during the time of slowest plant growth in your area's
grazing season. This may range from less than 36 to
more than 100 days, depending on conditions. For
example, while 36 to 42 days recovery time are adequate
for late summer and fall grazing in the northeastern
United States, a 90-day recovery period may be needed in
drier areas of other regions.

Remember, these are the total numbers of paddocks
needed when pasture plants grow the slowest. During
times of fastest growth, such as in May-June in the
Northeast, you will only need about 1/3 to 1/2 as many
paddocks, since the rest of the pasture area will be set
aside for conserving surplus forage.

Second, estimate how long the periods of occupation will
be, and decide whether to use one or two groups of
animals.

Third, use this equation to calculate the number of paddocks needed:

(recovery period ÷ occupation period) + no. of groups

For example, if 36-day recovery periods will be required, and you graze one group of animals with 12-hour (0.5 day) occupation periods, the number of paddocks needed will be: $(36 ÷ 0.5) + 1 = 73$.

With one group and 1-day occupation periods, the number of paddocks needed will be: $(36 ÷ 1) + 1 = 37$.

If one group grazes with 2-day occupation periods, then: $(36 ÷ 2) + 1 = 19$ paddocks needed.

If you divide your animals into two groups and graze within the same cell, just add one more paddock:

$(36 ÷ 0.5) + 2 = 74$; $(36 ÷ 1) + 2 = 38$; or $(36 ÷ 2) + 2 = 20$.

The more paddocks that can be formed to provide adequate recovery periods and short occupation periods, the better. Up to about 30 paddocks, each new division significantly shortens the average occupation period of all paddocks. Although the rate of benefit gain becomes less as more than 30 paddocks are formed, it's still worthwhile to make all the paddocks needed for adequate recovery periods.

Don't even attempt to use management intensive grazing with fewer than 10 paddocks, because the results will be very disappointing, and you may conclude that the method doesn't work. The more paddocks there are, the more flexibility exists for dealing with changing amounts of forage available during the season.

For example, if you only have six paddocks, and for some reason the forage production in one of them drops off, the decline would affect 1/6 of your pasture area. In contrast, if you have 37 paddocks, a problem with one of them only affects 1/37 of your pasture area.

--∞--

At this point, you may be thinking that building a lot of

paddocks will be too expensive, and that the management required to control grazing will take too much time and effort. By trying to answer likely questions that may come up and to provide you with enough information to get started doing it, I may be making it seem more complicated on paper than it actually is in practice. Once you're managing grazing correctly, you'll see that any costs that were involved in getting started will soon be paid back from the increased profitability of your farm. You'll realize that the method requires much less time, labor, and expense than feeding animals in confinement. Because of recent developments in fencing materials, the cost and building of paddocks will be surprisingly inexpensive and easy to do.

The most important point of all this is that you have enough paddocks to provide the required adequate recovery periods, with the periods of stay and occupation that you plan to use. The main idea is to graze the sward at the optimum stage of growth for the plants and animals. Your plan should be thought of as a starting point that you build on and change as you gain experience from observing results, and as the pasture improves. Don't try to force the plants and animals to fit a rigid schedule. Remember that these are guidelines to help you develop a management routine of planning, monitoring, controlling, and replanning if necessary.

PADDOCK SIZE AND FORAGE ALLOWANCE
After you decide how many paddocks to have, divide the total pasture area by the number of paddocks to roughly estimate the average area of each paddock. Paddocks don't have to be equal in area, but should produce more or less similar amounts of forage to facilitate moving animals on a fairly regular schedule, for your convenience. For example, pasture areas with fertile soils and good moisture conditions may produce twice as much forage as areas with poor, dry soils. Paddocks in the highly

productive areas should then be about half the size of those in the poor areas.

With electric fencing it's always possible and easy to relocate some fences or subdivide more if necessary. For example, as your pasture becomes more productive and the animals can't eat everything within the occupation period that you want to use (e.g. 12 hours), just divide the paddocks in half, in thirds, or whatever it takes to reduce the forage allowance enough.

As stocking density rises, there's more competition among the animals for feed and less selective grazing. Under heavy stocking density (e.g. 150 to 200 animal units/acre/12 hours), even high-producing dairy cows will graze uniformly and close to the ground. They will do well if forage quality is high, forage allowance and sward height are adequate, and their rations are correctly balanced with energy, protein, and mineral supplements needed for the desired production level.

Paddock sizes must be adjusted according to the intensity of management that's wanted. I can't tell you an exact paddock area to use, because that depends on how often you want to move your animals, on pasture productivity, on numbers, kinds, and sizes of the animals grazing, and their forage needs.

You can calculate what paddock areas should be to provide the needed amount of forage (forage allowance) from estimated livestock dry matter intake needs and pasture mass relationships. This can be done as frequently as every 12 hours to form portable-fenced paddocks (called "breaks" because they're broken out of a larger area) based on daily pasture mass measurements, or only once to estimate paddock size when you are planning the layout of permanent fences.

$$\text{Paddock area} = \frac{\text{(no. of animals} \times \text{dry matter intake)(days)}}{\text{pregrazing mass - postgrazing mass}}$$

You can get a close estimate of what pasture forage dry matter intake will be per animal per day, by calculating 3 to 4.5% of average bodyweight for dairy cows (depending on supplementation level), 3% for stocker cattle, 4% for lactating sheep, 4.5% for growing lambs, and 5.5% for lactating goats and growing kids. This assumes that your pasture is well managed and its forage is good to excellent quality.

For example, a herd of 70 Holsteins requiring 39 lb DM/head/day from pasture (calculated: 3% x average bodyweight of 1300 lb), are given a fresh paddock every 12 hours. Pregrazing pasture mass will be 2400 lb DM/acre, and postgrazing pasture mass will be 1400 lb DM/acre. (Dry cows and heifers will follow and graze down to 1000 lb DM/acre.) The paddock size to provide adequate forage allowance will be:

$$\text{paddock area} = \frac{(70 \text{ cows x } 39 \text{ lb DM/head/day})(.5 \text{ day})}{2400 \text{ lb DM/acre} - 1400 \text{ lb DM/acre}}$$

$$\text{paddock area} = \frac{70 \times 39 \times 0.5}{1000} = 1.4 \text{ acre}$$

Now, suppose you have a large rectangular paddock of 400 by 3000 feet, with permanent perimeter fence. You need to decide what size subdivisions (breaks) should be:

Breaks for 70 cows for 12 hours:

$400X = 1.4$ acres   1 acre = 43560 square feet
1.4 acres = 1.4(43560) = 60984 sq. feet

Solve for X, the unknown measurement of the breaks you need to make:

$X = 60984 \div 400 = 152$ feet

So, your breaks should be 152 x 400 ft for 70 cows for 12 hours. Of course you'll be able to see if the breaks are the right size, from your daily observation of how much forage the cows are leaving behind, how they behave (Do they act nervous as if they're hungry, or contented and relaxed like full cows?), and their milk yield. Having dry cows and heifers following to clean up behind the milkers makes grazing management easier. If the milkers leave too much forage behind, the followers can eat it and keep the sward in good condition.

The flexibility of portable fencing enables easy machine harvest of surplus forage, since break fences can be removed easily. Because conditions and situations vary so much, unless you do the above measurements and calculations, you'll just have to experiment with different sizes of paddocks.

Paddock size is important, but not nearly as important as being absolutely certain that you provide adequate recovery time between grazings.

EARLY SPRING MANAGEMENT
At least four problems must be dealt with in the first couple of spring rotations:
• the transition time needed by animals to adapt to eating green forage after months of eating relatively dry feed, to avoid scouring and bloat;
• keeping grasses in vegetative growth stages so they don't flower and go to seed (decreases their feeding value);
• the need to stagger the amount of forage available in paddocks so they don't all need to be grazed at once; and
• pugging of wet soils (not just a spring problem).

Scouring and Bloat
Pasture forage in early spring is extremely lush (only about 15% dry matter) and contains very high levels of nitrogen (crude protein content = about 30%). If you've been feeding protein supplement to dairy cows, you need

to stop feeding it or decrease the supplement level. Otherwise the animals have to eliminate excess nitrogen from their bodies and may scour besides. Eliminating nitrogen requires energy that could have been used for production. If you smell ammonia in your barn or milking parlor and cow poop is spraying you, you're probably losing money from decreased production and over feeding protein supplement.

Bloat can occur in pastured animals (mainly cattle) when they graze succulent legumes (alfalfa, red and white clover), especially in spring. Due to a complex interaction of animal, plant, and microbiological factors, a stable foam forms in the rumen. Because of the foam, gas can't escape. Pressure builds up in the rumen, causing it to swell and press against the lungs, preventing breathing. If not treated quickly, the animal dies.

Precautions are needed to avoid scouring, bloating, and death of your animals, and to allow time for microorganisms in animal digestive tracts to adapt to lush pasture forage. These are some things that can help:

1. During at least the first 2 weeks of spring turnout:
• Fill animals up with hay before putting them on pasture for a couple of hours late in the morning or early afternoon (first few days).
• Move animals to fresh paddocks only when dew or rain has dried off.
• Have dry hay always available to animals on pasture, in case some of them need additional dry fiber.
• Have water, salt, minerals, and sodium bentonite always available to animals on pasture.
• Watch animals closely; call a veterinarian at the first signs of bloating.

2. If you apply nitrogen fertilizer or manure to your pasture in early spring, be extremely careful because they result in forage being even more succulent. Before applying them make sure you'll be able to use all the forage that will result. It may be better to hold off applying them until early summer to boost forage production when you need it more. (see chapters 9 & 10)

3. Feed antifoaming agents (e.g. poloxalene, Bloat Guard) or add them to drinking water, beginning several days before the animals are first turned out to pasture, and continuing until they have adjusted to the lush forage.

4. Manage grazing so that the sward contains less than 50 percent clover or alfalfa. This can be achieved by allowing greater pregrazing pasture mass so grasses shade the legumes and decrease their amount in the sward.

5. Tendency to bloat seems to be hereditary, so the best long-term solution is to cull animals that bloat.

Keeping Grass Vegetative
Like all living organisms, including us, cool season grasses have an uncontrollable urge to reproduce, especially in the spring. But if grasses manage to flower and set seed, their forage production and quality greatly decreases during the rest of the growing season. Since this goes directly against your interests, your main management objective in spring must be to keep grasses in a vegetative stage of growth. There are several ways of achieving this; all require you to walk your pasture daily to observe what's happening:

Conserve Surplus Forage Set aside part (usually 1/2 to 2/3) of your pasture area for conserving surplus forage as hay or haylage. When plants regrow after machine harvesting to your target pregrazing mass, incorporate this area as needed into the grazing rotation.

This is my two preferred way of keeping all of the

pasture under tight control to maintain high sward density and forage quality. With this method you can manage paddock by paddock, keeping pre- and postgrazing masses within your target ranges. This is important to keep all plants short so that light conditions favor tillering and thickening of the sward, resulting in a vegetative, leafy stage of growth for subsequent grazings.

Increase Number Of Animals Bring in extra animals to graze until plant growth slows in midsummer, and remove some animals at that time. You can do this under a contract that either pays you for liveweight gain or on a per-head-per-month basis. If you charge by weight gain, animals have to be weighed when they enter and leave your pasture; during busy spring time this can be difficult to accomplish.

Rotate Fast Move animals to fresh paddocks more than once or twice a day, so all paddocks are top-grazed every 5 to 8 days. This lax grazing will result in clumpy pasture, but it won't cause a serious problem if done for a short time. As soon as plant growth rate slows, slow the movement of animals and graze paddocks closely again, allowing sunlight to reach the base of the sward. Your animals will be on a very high plane of nutrition, and production probably will increase, paying you for your extra work.

Use Big Breaks If you're break-grazing (subdividing large paddocks) with portable fencing, make large enough breaks so that all of the pasture is top-grazed every 5 to 8 days. When plant growth rate slows, return to small breaks sized according to forage allowance needs under close grazing management.

Set Stock (continuous graze) Open all paddock gates to allow animals to top-graze the entire pasture, leaving a high residual of about 2000 lb DM/acre (4-5 inches tall). This kind of management is useful on rough land where surplus forage can't be conserved by machine harvesting. This lax grazing results in tall, clumpy pasture that can thin the sward, reduce white clover content, and decrease digestibility, but is better than allowing most of the pasture to get out of control and go to stem and seedhead. When plant growth starts to slow down, begin closing paddock gates until you're again rotating through, one paddock or break at a time.

--∞--

No matter which of the above practices you use it's a good idea to provide hay free choice on pasture in early spring for animals that want it; this helps keep dry matter intake high and prevent bloat. Don't mow or harrow in spring; both spread parasites and manure, which makes forage less palatable.

Staggered Forage Accumulation
Graze Early

Graze some paddocks a little too early so that others are not grazed way too late. You'll have to begin grazing some paddocks when pasture mass reaches 1400 to 1600 lb DM/acre (2 to 3 inches tall). Make certain that plants have a fully developed green color in the first paddocks before beginning to graze.

Unless you start early in the first paddocks to be grazed, by the time the animals reach the last ones to be grazed in the first rotation (12 to 20 days later, depending on occupation periods), the forage in them will be too mature, and the animals won't graze them well. This isn't as much of a problem if you're willing and able to clip the last few paddocks soon after they're grazed. Otherwise, the forage that's left in the first rotation will be there to haunt you all season, and will decrease the productivity of those paddocks.

Stockpile Autumn Forage

Another way to stagger the amounts of forage available in paddocks is to stockpile (leave ungrazed) forage in autumn on part of your pasture. Although the nutritive value of this forage doesn't decrease much if it's frozen and covered with snow, palatability may be less; it's best used for non-producing livestock. Grazing can begin very early (e.g. mid-March) on stockpiled areas. Because soils at this time of spring usually become wet and soft, very careful management is needed to avoid damaging soils and swards (see management under Wet Soil Conditions at the end of this chapter).

LATE SPRING MANAGEMENT

When the weather begins to get hotter with approaching summer (e.g. early June), it may be beneficial to mow your pasture with a mower set high to remove just the stems and seedheads that got by your early spring management. The stems and seedheads bother your livestock's eyes (possibly linked to pink eye), and make surrounding forage less palatable and nutritious.

Nitrogen Fertilizer/Manure

If you need to boost forage accumulation to get more easily through a summer slump period of plant growth, late spring to early summer is a good time to apply manure or nitrogen fertilizer. Either apply liquid manure (about 3000 gallons/acre), other manure (composted or uncomposted) pulverized and spread thinly so plants aren't smothered, or nitrogen fertilizer (don't use urea: it kills earthworms). Try to apply any of these materials right after animals graze a paddock, and if possible before a rainfall (see chapter 10).

WET SOIL CONDITIONS

Wet soils complicate grazing management, but if pugging from animal hooves isn't severe, plant production won't decrease much if at all; it may even increase because short grazing periods with some pugging stimulates pasture plant growth. If pugging is very severe, plant production will decrease for a few weeks to a few months, but usually becomes better than before the pugging.

Farmers have found these ways of dealing with wet soil conditions that don't severely damage the pasture:

- Use short grazing periods of only 2 to 3 hours during the day and again in the evening. Remove them from the pasture as soon as they stop grazing, and keep them on a concrete pad, in the barn, or on a well-drained area. Most pugging occurs when animals have finished grazing and are just walking. Don't leave animals on wet pasture soil overnight. Provide hay, haylage, or silage, and any supplements free choice to animals after they graze, not before. This ensures that they're hungry when they're on pasture, so they can quickly eat their fill.

- Soil is dried mainly by plant evapotranspiration, and tall plants transpire more water than short ones. If you allow plants to grow to about 6 to 8 inches tall before grazing down to a short residue (1-2 inches), the soil will dry quicker than if the plants are grazed when they reach 4 to 6 inches tall.

- Allocate only the amount of forage needed (taking into account the need to leave a tall residue) and the paddock area that contains it for the short grazing periods. This reduces walking and pugging.

- Graze back ends of paddocks first. You'll have to push the animals to the back, walking over ungrazed plants.

- Always backfence so that animals can't return to areas already grazed.

- Keep the animals moving slowly through the rotation by allocating a little of the pasture area at a time, so that only part of the pasture gets pugged. This is difficult to do in the spring when plant growth rate is very fast, and you want to be grazing through fast to control the growth and keep the plants vegetative. Later when soils are drier you can deal with the pasture areas that got out of control.

- Graze poorly drained paddocks during drier weather, and well-drained paddocks during wet weather.
- Construct all-weather lanes (see chapter 6).
- Provide more than one entrance to paddocks that allow animals to fan out quickly, distributing themselves throughout the paddock.
- Use different gates for animals to enter and leave each paddock.
- Grade paddocks that tend to be wet into humps and hollows to drain surface water quickly into open drains. Leave 30 to 40 feet between hollows. Hollows need to be cleaned every year with a drain spinner.
- Spread gravel in paddock entrances and around water tanks to prevent mud being carried into paddocks and soiling the forage.
- Broadcast seed on severely pugged areas right after grazing. This is an opportunity to introduce new varieties of grasses and legumes. Birdsfoot trefoil and annual ryegrass reportedly establish quickly on such areas.

DROUGHT CONDITIONS
Depending on your plants, there are at least two (seemingly contradictory) ways of managing pasture during drought:

- <u>Short grasses and legumes</u> (e.g. Kentucky bluegrass, perennial ryegrass, orchardgrass, timothy, white clover). The management used with these plants is based on the concept that soil is dried mainly by plant evapotranspiration, so plants that are short (4-6 inches tall) when grazing begins, use less water and dry soil slower than tall plants (taller than 6 inches). Recovery periods must be long enough for plants to reach an average height of 4-6 inches before grazing again.

Leaving a tall residue (more than 2 inches) after grazing dries soil faster than a short residue. A tall residue usually should not be left, no matter what the soil moisture conditions are. Tall residue generally results in poor quality forage during the rest of the season. Plants should be left short (less than 2 inches tall) so that regrowth will be high-quality leafy material.

• <u>Tall grasses and legumes</u> (e.g. bromegrass, quackgrass, reed canarygrass, red clover, alfalfa, birdsfoot trefoil).
The management used with these plants is based on the concept that the soil needs to be kept covered and shaded by a living mulch to prevent it from drying out. Plants are allowed to grow 10 to 16 inches tall before being grazed down to about 6 inches tall. This management results in thinner swards and stemmier forage that is less digestible, but the lax grazing allows livestock to compensate by grazing more selectively.

ROTATIONAL SEQUENCE

If you graze paddocks in the same sequence every year, you will soon have as many different pasture plant populations as you have paddocks. This is because some paddocks will always be grazed too early for certain plants, some will be grazed just right, and others will be grazed too late. The way to distribute the stresses of early and late grazing in the first rotation, is to start with a different paddock each year and use the other paddocks in a different sequence, if possible.

During wet spring and fall conditions you will have to graze higher and drier areas to avoid problems with mud and soil pugging, so it may be impossible to make much of a change in the sequence that your paddocks are grazed. But even if you alternate the first grazing each year between only two paddocks, it's better than not changing the sequence at all. As always, do what you can, knowing what the ideal situation would be.

Later in the season, if you notice that the forage in a paddock is ready to be grazed ahead of its time in the sequence, go ahead and graze it. Grazing out of sequence won't damage the plants, because their faster growth indicates that conditions in the paddock are more favorable than in other areas of the pasture, and that they have had adequate time to recover under those conditions. Grazing management must be flexible, to deal with the ever-changing pasture environment.

REFERENCES

Appleton, M. 1986. Advances in sheep grazing systems. In J. Frame (ed.) *Grazing*. British Grassland Society, Berkshire, England. 250 p.

Bingham, S., and A. Savory. 1990. *Holistic Resource Management Workbook.* Island Press, Washington, D.C. 183 p.

Bircham, J.S., and C.J. Korte. 1984. Principles of herbage production. In A.M. Fordyce (ed.) *Pasture: The Export Earner.* New Zealand Institute Agricultural Science, Wellington, New Zealand.

Blaser, R.E., R.C. Hammes, Jr., J.P. Fontenot, H.T. Bryant, C.E. Polan, D.D. Wolf, F.S. McClaugherty, R.G. Kline, and J.S. Moore. 1986. *Forage-Animal Management Systems.* Virginia Polytechnic Institute and State University, Blacksburg, Virginia.

Fletcher, N.H. 1982. *Simplified Theoretical Analysis of the Pasture Meter Sensing Probe.* Australian Animal Research Laboratory Technical Paper. Commonwealth Scientific and Industrial Research Organization, East Melbourne, Australia.

Foster, L. 1996. Wet and dry pasture management. *Stockman Grass Farmer.* 53(10):5-6.

Frame, J. 1981. Herbage mass. In J. Hodgson, R.D. Baker, Alison Davies, A.S. Laidlaw, and J.D. Leaver (eds.) *Sward Measurement Handbook.* British Grassland Society, Berkshire, England.

Griggs, T.C., and W.C. Stringer. 1988. Prediction of alfalfa herbage mass using sward height, ground cover, and disk technique. *Agronomy Journal.* 80:204-8.

Hintz, R. 1993. The causes and prevention of alfalfa pasture bloat. *The Haymaker*, Beachley-Hardy Seed Co., . Summer. p. 5, 7.

Holmes, C.W., G.F. Wilson, D.D.S. MacKenzie, D.S. Flux, I.M. Brookes, and A.W.F. Davey. 1984. *Milk Production From Pasture.* Butterworths. Wellington, New Zealand.

Korte, C.J., A.C.P. Chu, and T.R.O. Field. 1987. Pasture production. p 7-20. In A.M. Nicol (ed.) *Feeding Livestock on Pasture.* New Zealand Society of Animal Production, Hamilton.

Nation, A. 1992. Al's Observations. *Stockman Grass Farmer.* 49(7):1, 21.

Nation, A. 1992. Winter pugging. Allan's Observations. *Stockman Grass Farmer.* 49(10):20-21.

Nation, A. 1992. Wet weather grazing tips. *Stockman Grass Farmer.* 49(12):17.

Nation, A. 1993. Spring pasture management similar to flying a jet. *Stockman Grass Farmer.* 50(4):1, 8.

Nation, A. 1994. Pugging damage small if grazing time is short. *Stockman Grass Farmer.* 51(8):5.

Nation, A. 1995. Wet weather grazing tips. *Stockman Grass Farmer.* 52(12):13.

Nolan, T., and J. Connolly. 1977. Mixed stocking by sheep and steers: A review. *Herbage Abstracts.* 47(11): 367-74.

Parsons, A.J., and I.R. Johnson. 1986. The physiology of grass growth under grazing. In J. Frame (ed.) *Grazing.* British Grassland Society, Animal and Grassland Research Institute, Berkshire, England. 250 p.

Rayburn, E. 1988. The Seneca Trail Pasture Plate for Estimating Forage Yield. Seneca Trail Research and Development, Franklinville, New York. Mimeo. 5 p.

Reid, R.L., and G.A. Jung. 1973. Forage-animal stresses. p. 639-653. In. M.E. Heath, D.S. Metcalfe, and R.F. Barnes (eds.) *Forages.* Iowa State University Press, Ames.

Savory, A. 1988. *Holistic Resource Management.* Island Press, Washington, D.C. 564 p.

Sheath, G.W., R.J.M. Hay, and K.H. Giles. 1987. Managing pastures for grazing animals. p. 65-74. In A.M. Nicol (ed.) *Feeding Livestock on Pasture.* New Zealand Society of Animal Production, Hamilton.

Smetham, M.L. 1973. Grazing management. In R.H.M. Langer (ed.) *Pastures and Pasture Plants.* A.H. & A.W. Reed, Wellington, New Zealand.

Smith, B., P.S. Leung, and G. Love. 1986. *Intensive Grazing Management.* Graziers Hui, Kamuela, Hawaii. 350 p.

Smith, B. 1995. Variable density grazing. *Stockman Grass Far;mer.* 52(1):1,4-6.

Voisin, A. 1959. *Grass Productivity.* Philosophical Library, New York. Reprinted in 1988 by Island Press, Washington, D.C. 353 p.

Voisin, A. 1960. *Better Grassland Sward.* Crosby Lockwood & Son, London. 341 p.

6
Paddock Layout, Water, & Fencing

Over the river, the shining moon;
In the pine trees, sighing winds;
All night long so tranquil -- why?
And for whom?

Hsuanchueh

Right from the start, try to design an ideal paddock layout that requires minimal work to use and is as intensive as possible. You can use a less-intensive level of grazing management while the improved productivity pays for the rest of the setting-up costs. You'll always be able to build toward the ideal layout, if you design it that way in the beginning.

You'll need to answer several questions: How intensive do you want your grazing management eventually to be? How much are you willing to spend in the first year to start the system? Which areas if any will be set aside in spring for machine harvesting? Which animals will need to return to a central handling facility (e.g. milking shed or parlor), and how convenient must it be? Will you be able to provide drinking water to all of the paddocks? How will you get animals through or around low wet areas? How will you deal with poorly drained pasture areas?

Keep the design flexible for changes in production level and adding other kinds of livestock. In planning the layout, remember that fences cost less, are easier and faster to build, and last longer when built in straight lines.

TOPOGRAPHY AND HANDLING FACILITIES
The shape of paddocks and where lanes and dividing fences will be placed depends a lot on the topography of your pastureland, and its location relative to the barn or

other handling facilities. This is especially important with dairy animals that have to be milked twice a day. Ideally, if you are a dairy farmer, you should plan for future placement of a milking shed in the center of your pastureland.

Use an aerial photograph of your farm if possible, or make a sketch of it, to design the paddock layout. It's much easier to make changes if you draw paddocks and lanes on paper, before building any fences.

Place permanent fences on keylines (points where ridges change to hillsides and hillsides change to valleys or terraces). This is important because large differences in soils and microclimates exist on either side of these points.

Try to form paddocks that are more square-shaped, rather than long and very narrow. This is because less fencing materials are needed to fence in a square-shaped paddock than a long narrow one, even though they contain the same area. (Here's a chance to use your high school geometry!) If your pastureland is level-to-rolling, though, you can divide it into large permanent rectangles that are subdivided as needed with portable fencing.

Rough, Hilly Land
If your land is rough and hilly, as most permanent pastures tend to be, you should take slope aspect and location of hill crests into consideration when designing the paddock layout. Try to enclose as similar an area as possible in one paddock. Water relationships, aspect, soil type and depth, and vegetation vary greatly depending on topography. For example, land with north-facing slopes should be in paddocks separate from land with south-facing slopes, if the areas involved are large enough to make a difference and to form separate paddocks.

South-facing soils warm up quicker in spring, get hotter and dryer in summer, and stay warmer longer in fall than north-facing soils. So in early spring, south-

facing slopes may be ready to graze 2 to 3 weeks before
north-facing slopes. In midsummer, plants on south-
facing slopes may grow very slowly because of the hot, dry
conditions, while plants on north-facing slopes continue
to grow rapidly. Plants on south-facing slopes can be
grazed longer in fall than plants on north-facing slopes.
Proper management of such areas is possible only if they
are in separate paddocks.

East-facing slopes are more similar to north-facing
than south-facing slopes. West-facing slopes are more
similar to south-facing slopes.

Hill crests also should be contained in separate
paddocks to promote uniform grazing and decrease
nutrient transfer to favorite camp sites (see chapter 4).
Paddock divisions along hill crests force animals to graze
sunny and shaded aspects of the hills, and prevent them
from camping on sunny slopes after grazing the shaded
slopes. (Figure 6-1)

Level-To-Rolling Land

If your pastureland is level-to-rolling, it's not necessary to
permanently divide it into small paddocks. Instead, you
can fence in large, long, wide rectangular areas with
permanent perimeter fencing. Inside use portable fencing
to subdivide into fresh areas (breaks) of forage that range
from square- to triangular-shaped. This way of dividing
pastures saves a lot of money in fencing materials, and
makes it easy to set aside large areas of the pasture for
machine harvesting of surplus forage. (Figures 6-1-5)

For cattle and horses, build permanent one- or two-strand
high- or low-tensile fences around large rectangular
paddocks up to about 900 feet wide and as long as you
want. Subdivide the rectangles across their widths by
using a polywire or polytape in front of the animals,
supported either on light portable fiberglass or plastic
posts or on rolling fence (e.g. Gallagher tumblewheels).

Tumblewheels can be used to subdivide rectangles up to about 600 feet wide. If portable posts are used, the rectangle shouldn't be more than 900 feet wide for easy moving of front and back fences.

If a permanent lane doesn't border the large paddock, follow animals with a polywire L-shaped back fence, also supported on portable posts, or on tumblewheels. The back fence prevents animals from eating regrowth, and the L-shape forms a lane for them to go to the barn or to water, if a water trough is not moved along with them. The lane formed by the back fence can be alternated each rotation between the two sides of a large paddock, so that one side doesn't get too worn and compacted or muddy. This capability of changing the lane location is especially important during wet spring and fall conditions.

If a permanent lane borders the large paddock, a straight back fence is used, since livestock can exit breaks through gates spaced along the lane.

Move the front polywire to give the animals only enough forage for the occupation period that you want to use. The advantage of using tumblewheels is that one person can very quickly give animals a triangular-shaped break of fresh pasture by moving every other end of the polywire at each setting. (Square- or rectangular-shaped breaks can be given with tumblewheels by moving both ends of the dividing wire.) Reels at each end of the tumblewheels make moves easier, because you can let out or wind up the polywire as needed. Another way is to use a reel with a friction brake at one end; this allows you to pull out the polywire as needed.

A front polywire supported on portable posts also works well, but requires a little more time and walking to move it. Portable posts that have a lip for stepping them in enable faster moving of the fences. One way to move a front polywire on posts is to move one end first, and then reposition the posts to where you want them, without taking the polywire off the posts. Then just take up the slack with a reel at one end.

Both tumblewheels and polywires on portable posts can be moved without turning the perimeter fence off. It's a good idea to disconnect the polywire/post front fence from the electricity, though, because it's tricky to move it without touching it. Clips are available for disconnecting polywires from perimeter fences.

The back fence doesn't have to be moved every time the front wire is moved, but the more frequently the back fence is moved the better. This is because the longer the animals have access to a pasture area, the longer the occupation period is for that area. Occupation periods that are too long may allow grazing of regrowth in the same rotation, and the pasture will not reach its full potential. Short occupation periods also reduce treading damage when soils are wet. Moving front and back fences every 12 hours results in highest plant and animal productivity. Whatever you do, move the back fence frequently enough so that regrowth can't be grazed in the same rotation.

John Clark is a Vermont dairy farmer who could win a prize for simplifying subdivision breaks. He uses fiberglass posts and three separate strands of polywire, each with its own reel, to create two breaks. The front and middle wires form a break for milking cows. The middle and back wire form a break for following dry cows and heifers. When he removes the cows from the front break for milking, he allows the dry cows and heifers to enter it. Then he reels up the back wire, and moves it around to the front to form a fresh break for the cows when they return from milking. A portable water trough is moved with each group of animals.

For sheep and goats, build rectangles that are multiples of 145 feet wide (e.g 145, 290, 435, 580, etc.) by as long as you want. The 145-foot increments are needed because the electric net fencing for small livestock usually comes in 150-foot rolls. So why not make the rectangles in 150-foot increments? Because the net fencing can shrink after a few years, and then it won't reach across the rectangles. (Premier Fence Systems now has available electric net fencing in lengths ranging from 15 to 600 feet.)

Suppose, for example, that you build a 5-strand permanent high tensile fence (more on this later) around a 580- by 1500-foot area. Then you can subdivide it lengthwise into four, 145- by 1500-foot smaller rectangles, using portable posts and three strands of polywire. (Step-in posts that have clips at correct positions are available. Multiple reel assemblies and single reel, multiple polywire combinations are also available for quickly and easily positioning three or more strands of polywire.) Then net fencing can be used to subdivide the 145-foot wide rectangles. Three sections of net fencing should be used for each group of animals: one section in front, another in back, and the third ahead of them to limit the next area that they will move into.

Keep records of when and where portable fencing was placed, and relocate the fencing at the same positions every rotation. This is to be certain that all areas of the pasture receive adequate recovery periods. If you don't relocate the fencing at the same positions in every rotation, some areas may get less recovery time than they need, and start you into untoward acceleration. When using portable front fencing, tie a short piece of twine or surveyors' tape to the perimeter fence at each spot where you connect the front wire(s). Relocating net fencing at the same positions is fairly easy, because of the clear checkerboard effect on the pasture that results from using this method for grazing sheep or goats.

LANES

You must be able to put animals in paddocks where and needed. These are aspects of building all-weather access lanes that are needed to manage pasture well:
• Energy is used in walking that otherwise could be used

in production. For example, milk yield per cow decreases about 2 pounds per mile walked, and somatic cell count increases with walking. So lanes should be as short and as direct as possible, to allow animals to reach all paddocks with the least amount of walking.

- Make lanes only as wide as needed, because forage in lanes usually becomes soiled and wasted. Lanes should be just wide enough to allow the animals and necessary machinery to get through. Animals will conform to lane widths, and walk single-file if they aren't rushed. For example, 6- to 12-foot wide lanes work well for animals, but aren't wide enough for large machinery. Lanes 14- to 24-feet wide are needed for tractors with mowers, rakes, balers, and wagons.

- Avoid right-angles in lanes, because animals dislike walking around them. Walking around tight bends also damages animal hooves and lane surfaces.

- Locate lanes on the highest and driest ground, following tops of hills, if possible. Use culverts and gravel fill if necessary to make certain that your animals can walk through the lanes without dropping out of sight in mud. This is especially important in the earliest part of the season, when grazing must begin so that the forage doesn't become too mature, and during wet fall conditions.

- It's best to have lanes cross waterways and streams at a right angle; these crossings should be reinforced. This avoids damage to stream banks, sedimentation of streams, and having livestock in mud.

- In low areas place culverts at least 18 inches below ground level, so they aren't crushed by animal or machine traffic. Angle culverts so water runs through quickly and doesn't leave silt in the culverts. Dig holes in front of entrance openings for silt collection, and periodically clean them so silt doesn't fill the culverts.

- Fabrics (e.g. Mirafi 500) are available that can be placed over muddy spots before covering with gravel. This keeps mud from mixing with the gravel, and forms a stable surface over wet areas in lanes and around drinking tanks. The fabric and gravel fill should be put in when the areas are dry, for best results.

- In lane areas that are less wet, a thin layer of 3-inch crushed stone helps support animals and machinery.
- Depending on soils and herd size, reinforce at least 500 to 1000 feet of the lane nearest the milking facility.
- The main idea in improving lanes is to shape them so that water drains off. This is achieved by crowning the lane 6 to 12 inches high at center for drainage; be sure to maintain the crown by grading periodically.
- Never use gravel-size stone on lane surfaces because it damages hooves.
- Follow conservation service recommendations for reinforcing lanes and crossing streams.

Wisconsin dairy farmer Charlie Opitz has developed a way of stabilizing lanes that allows a lot of cows (1200 in herds of 200 each) to pass back and forth twice a day regardless of weather. First he grades the lane level to keep dump trucks level when spreading crushed stone. Then he spreads 15 yards of 1.5- to 2-inch crushed stone per 130 feet of 12-foot wide lane. On top of this he spreads 15 yards of ground limestone per 175 feet of lane. This currently costs about $2.00 per foot of lane.

An ideal lane and gate arrangement has gates in paddock division fences and at both paddock corners in V-shaped openings on the lane (Figures 6-1, 6-2). This design allows animals to move in any direction in the lane or to adjacent paddocks. If you can't build the ideal setup, be sure to place lane gates in the corner of paddocks nearest the barn, milking shed, and/or water. Otherwise, you may have some animals that can't figure out that they need to walk away from the barn to go through a gate into the lane before going to the barn. If this happens, instead of all of your cows coming to be milked, you'll have to go back to the paddock and chase out the few cows that get trapped in the paddock corner nearest the barn. This is especially a problem with heifers, and when you're just beginning to manage grazing, because animals take a few days to learn what they're supposed to do.

If animals don't have to leave the pasture for drinking water or to be milked, you can put gates in paddock division fences, and move animals to fresh paddocks without entering the lane. The lane can then be used only once per rotation for returning animals to the first paddock to start the next rotation, and can be grazed as a paddock itself, since the forage in it wouldn't become soiled by animals moving through it.

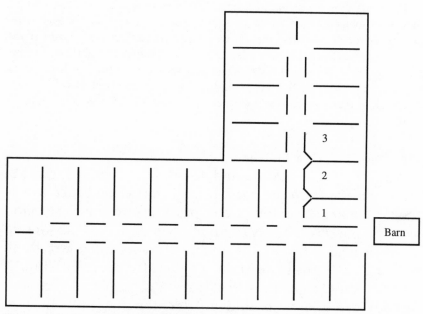

Figure 6-1. Example of a paddock and lane layout on level-to-rolling land, with gate openings allowing access to all paddocks and lane from all directions. Paddocks 1, 2, and 3 show V-shaped openings with gates closed. (Courtesy of Gallagher Power Fence)

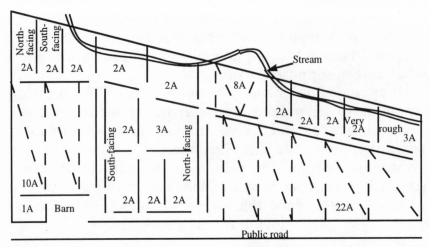

Figure 6-2. An example of paddock and lane layout on rolling-to-rough hilly permanent pastureland grazed by dairy cows. The 8-, 10-, and 22-acre paddocks are set aside in spring for machine harvesting. After adequate recovery time, these three large paddocks are grazed using portable fencing (polywire/post or tumblewheels) to subdivide them. Portable fence is moved every 12 hours, as indicated by --- lines. Back fence is installed in the large paddocks after the first two 12-hour grazings, and is then moved every 24 hours (see Figure 6-3 for back fence detail).

DAY/NIGHT PASTURES

If some of your pastureland for dairy cows is located a mile or more from the barn, it might be a good idea to use the land that's far away as day-pasture, and the closer land as night-pasture. This is so your cows don't take so long to come to the barn in the morning; you can start them walking well ahead of time for the afternoon milking. If you use day/night pastures, just be certain that recovery periods are adequate, and not too long, within the two areas. You can still use short (12-hour) occupation periods, but you may have to graze additional animals or machine-harvest more surplus forage in each area. This is because rotations within each area will take twice as long as they otherwise would have if grazing was all in one area before starting in the other.

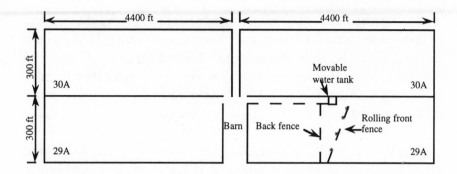

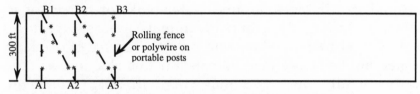

First paddock is about 50 ft wide; for second paddock move A1 about
100 ft to A2; for third paddock move B1 about 100 ft to B2, and so on.

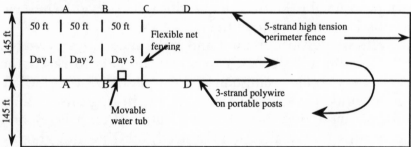

Move one section of flexible net fence once a day to give just enough
forage for 24 hours. When animals are between sections B and C,
move A to D, and so on, rotating around the large rectangles.

Figure 6-3. Examples of using large rectangular paddocks subdivided
with portable fencing on level-to-rolling pastureland. (Not drawn to
scale.)

DRINKING WATER

The source and distribution (reticulation) of drinking
water for a pasture area is an extremely important

consideration in planning paddock layout. Water is the nutrient needed in largest amounts by livestock, so make certain that an adequate supply of good quality, clean, cool water is available at all times. If you wouldn't drink from the water source, don't force your animals to drink from it. (see chapter 7)

For highest plant and animal production, it's best to provide water in all paddocks, even though only one group may be grazing a pasture. If water is near where the animals are grazing, they need to walk less to drink, and nutrients from manure remain in paddocks where they're needed, not in lanes, around water tanks or ponds, or in streams. When one group of animals grazes a pasture, you can get by for awhile with only one source of drinking water, but be aware that animals must be able to reach the water at all times, and that you will not achieve the highest levels of plant and animal production. When two groups of animals graze the same pasture area, water must be provided in each paddock, since one of the groups must be kept in a paddock to prevent the two groups from mixing together.

With one group grazing and only one source of water, it is possible to use lanes to allow animals to reach the water. Don't build triangular-shaped paddocks that radiate out around a central water supply, in regions having reliable, well distributed precipitation (nonbrittle) and moist soils (especially fine-textured) during the grazing season. This is a good design that enables easy handling of animals, but in the small paddocks usually needed on farms in humid regions it results in transfer of nutrients from wider parts of paddocks to narrow parts near the water supply, and compaction of moist soil in narrow areas. This is because animals (especially dairy cows) prefer to be near their water supply, and will graze in the wide areas but camp in the narrow ones. Extremely high levels of nutrients can build up in the narrow areas, while nutrient deficiencies develop in wide parts. This nutrient transfer combined with soil compaction causes the pasture plant community to deteriorate. In arid regions, where soils are drier and paddocks are large, the advantages of having one water source at the center of radiating triangular-shaped paddocks outweigh any disadvantages.

Water Source And Supply

Traditionally, animals have drunk directly from streams, rivers, ponds, and springs, usually standing in the water to drink or cool off. In humid areas, this practice is no longer possible, because of the severe damage it causes to water quality and stream and river banks.

Except for convenience, it never was a good idea even for livestock production. Water quality for the livestock drops drastically as soon as they enter the water source. The first animals to drink (dominants) get relatively good quality water. The following animals (submissives) get progressively worse quality water as sediments are stirred up and the animals poop and pee. Submissive animals may not be able to drink at all. There have been incidences of submissive animals dying because they became so full of mud from their drinking water! Besides the mud and manure that they take in, livestock that drink directly from surface water are susceptible to infection by water-borne pathogens and parasites. Imagine what this kind of situation does to production levels!

Stream And River Bank Fencing

Fencing off stream and river banks is needed in nonbrittle environments to improve quality of water flowing through farms. This directly benefits the farm in terms of better quality drinking water for livestock and increased plant and wildlife diversity along the waterways. It also benefits everyone downstream by reducing the load of nutrients and sediment from manure and bank erosion that the water carries. Federal and state agencies encourage fencing of stream and river banks, and will provide technical and financial assistance for doing it. Ask your local NRCS and Extension people for help in fencing and providing alternative ways for your livestock to drink.

Stream and river bank fencing provides many benefits:

- Stabilizes banks and reduces erosion. By volume, sediment causes the most pollution of surface waters. Much of the sediment comes from soil erosion along stream and river banks that are grazed. After fencing banks in a nonbrittle environment, plant species diversity along the banks increases. The new vegetation binds soil particles and absorbs some of the water's force, resisting erosion and collapse of the banks. Limiting access of livestock to stream and river banks prevents water pollution, soil erosion, and loss of productive land.
- Improves water quality. Runoff from cultivated fields, pastures, and feedlots pollutes surface water by washing soil, fertilizers, pesticides, and manure into nearby streams. Dense vegetation along fenced streams and rivers traps sediment, pesticides, and nutrients before they reach the water.
- Protects livestock health. Keeping livestock out of surface water reduces contact with pathogens that cause animal diseases (e.g. bovine leptospirosis, mastitis, lungworms) which decrease production and increase costs.
- Provides habitat for birds and other wildlife. Vegetation along streams and rivers provides food, cover, and nesting sites for 80 species of birds (e.g. herons, egrets, bluebirds, belted kingfishers, mallards, and pheasants) and many small mammals. Wildlife diversity decreases pests (e.g. flies) of livestock grazing nearby).
- Improves fish habitat. Vegetation along stream and river banks improves water quality, provides protective cover, and increases food supply for fish. Leaves and insects that fall into the water from the vegetation increase the amount and diversity of aquatic life in the water. Shrubs

and trees along the banks shade the water and keep it cool. This attracts fish and creates pleasant places for you to fish and swim during the summer, while your animals are feeding themselves by grazing elsewhere.

- Enhances the landscape. Excluding livestock results in a profusion of wildflowers and shrubs that attract birds and mammals, adding a variety of shapes, colors, sounds, and activity that add greatly to beautifying farm landscapes and improving your family's quality of life.
- Creates good neighbors. Everyone shares water. Improving surface water conditions on your farm helps everyone downstream. Your example may prompt your neighbors upstream to protect surface water flowing through their farms. We are all neighbors and all downstream.

Water Pumps And Gravity Flow
Since livestock should no longer drink directly from surface waters, we must find ways of delivering water to livestock. Fortunately, several excellent ways already exist:
- Ponds and springs can be developed to provide drinking water through gravity flow to pastureland below the water source. Extension and NRCS people can provide information and assistance you need to accomplish this.
- Pumps needing no electricity or motors are available to lift and pressurize surface water for distribution to livestock. Water rams use the force of a small drop in water level to lift water from 6 to 500 feet. Sling pumps lift water using the force of stream flow to lift water up to 82 vertical feet and deliver it up to 1 mile away. Pasture nose pumps enable animals to pump their own drinking water from surface water or shallow well, lifting it from a depth of up to 26 feet and moving it up to 126 feet horizontally. For information, contact the Rife Hydraulic Engine Manufacturing Co., P. O. Box 857PI, Montgomeryville, PA 18936; Phone: 800-RIFE-RAM.
- Water can be lifted from shallow wells using windmills,

available from O'Brock Windmill Distributors, 9435 12th
St., North Benton, OH 44449.
- A solar powered pump is available from Apollo Energy
  Systems of Navasota, Texas (Phone: 800-535-8588). This
  pump can transfer water from surface waters into tanks
  for livestock to drink. The unit has enough power to
  operate an energizer as well as pump water.
- Water lifted from surface water or shallow well with
  one of the above pumps or with a fuel-powered pump
  can be stored in large tanks on towers or high points of
  land, and distributed throughout the pasture by gravity
  flow. Another way is to deliver the water to grazing
  livestock with a large tank on a trailer or truck that is
  moved with the animals. Tanks that have been
  damaged too much for milk transport, but suitable for
  your needs may be available from dairy cooperatives.
  Animals can drink from drinking cups attached around
  the tank, or from a trough(s) with full-flow valves
  attached to the tank with 3/4-inch plastic tubing.

Water Reticulation

If your water supply is pressurized by pumping or gravity
flow, you can use black or white plastic tubing to carry
water to all paddocks. White tubing can be laid above
ground, and water in it doesn't get quite as hot as in black
tubing. It's best to bury black tubing, unless it will be very
well shaded by plants or something else such as rock
walls, to keep water cool enough. A telephone cable plow
can be used to bury tubing.

To make sure that water flow is adequate, use at least
3/4- or 1-inch tubing for up to 100 cows or similar amount
of animal units. For 100 to 200 cows use 1 1/4-inch tubing,
and 1 1/2-inch tubing for more than 200 cows. It's best to
install a complete loop of tubing so that water can flow
from the source in two directions to any point.

During hot weather check the water temperature in
drinking tanks. We have found that water coming

through long runs of black tubing lying on the soil surface, even though covered with plants, can become too hot to touch, let alone drink.

During autumn when freezing temperatures occur at night, water in above-ground white tubing freezes and doesn't thaw during the day; water in above-ground black tubing freezes at night, but thaws during the day.

Black tubing is available from hardware and farm supply stores, and fencing dealers, who also can supply full-flow valves (e.g. Jobe, Maxi Flow, or Hudson) and quick couplers. White tubing, valves, and quick couplers are available from Kentucky Graziers Supply, 1929 S. Main St., Paris, KY 40361; 800-729-0592.

Above-ground tubing can be drained in the fall by opening both ends and all faucets, picking it up at one end, and walking to the other end while holding the tubing up and moving it hand over hand. Then it can be left out during winter without being damaged.

If you bury tubing, make certain that it can be drained in the fall by placing drain taps at low points and breather taps at high points.

Number and size of water troughs needed depends on number and size of animals. For example, one 300-gallon trough per paddock is adequate for up to 150 dairy cows; for larger herds, two troughs are better. If you use two troughs, place one near the front of paddocks and one 2/3 of the way to the back. This decreases the distance animals have to walk, fertility transfer within the paddock, and pugging and treading damage.

Farmers have found problems with placing one trough under every other paddock division fence to serve two paddocks with one trough: 1) it allows animals only one-fourth the access to water that they otherwise would have had with an entire trough; 2) one dominant animal can prevent others from drinking; and 3) the fence may electrify the water and decrease or stop drinking.

Water can be supplied to paddocks from the tubing in

at least two ways:

- Make connections to full-flow float valves on 5- to 300-gallon troughs permanently located in paddocks.
- Place faucet or quick-coupler taps for garden hoses in the tubing at intervals (e.g. every 100 feet) that will allow you to use a garden hose to connect to a full-flow float valve for a trough that you move along with the animals. If tubing is placed along a central lane, four or more paddocks or breaks can be serviced from each tap.

Small troughs are easy to empty and move by hand, and you can place them where you want to increase soil fertility from manure. Lightweight troughs are available from farm supply and fencing dealers, or you can cut used 55-gallon plastic barrels in half. It's a good way to reuse barrels, but be careful with them because livestock can tip them over easily when empty, due their small diameter.

Python livestock watering equipment is by far the best that I've used. The large-diameter (32-inch) trough holds 35 gallons and is almost impossible for livestock to tip over, even when empty. Its low side (16-inch) enables most livestock to drink from it. Trough and full-flow valve weigh only 22 lb and can be dragged easily from paddock to paddock with the connecting lead tube. Current (1997) price of trough and valve is $240; more expensive than plastic half barrels, but it can make your life a lot easier. Python is sold by Kiwi Fence, 1145 E. Roy Furman Highway, Waynesburg, PA 15370; 412-627-8158/9791 fax. The equipment is extremely simple, but pay attention to instructions for connecting the trough lead to hydrants: push, turn, *relax* , pull. Relaxing your hold on the lead before pulling is important. Also, be careful not to push water line tubing so far into hydrant ends that it blocks movement of hydrant valves; tubing doesn't have to go in any farther than the coupling threads.

If there's good water pressure and you use a full-flow valve, one 30- to 35-gallon trough can work for up to about 70 cows. For more cows use a larger trough, or two or more small troughs separated by enough distance so that cows can easily use all of them at once. Whatever size trough you use, it needs to be full of water before animals enter a paddock; otherwise the sound of water running into an empty trough results in many or all animals wanting to drink at once. If this happens, they will drink all the water and probably tip the trough over. Be especially careful when using small trough to make certain that livestock water needs are met.

Portable troughs are especially useful for providing water to animals grazing within portable-fenced breaks of large paddocks. It's best to move troughs along with livestock every time they're moved to fresh breaks or paddocks. A benefit of moving a water trough with animals is that smaller troughs can be used, since all of the animals don't need or want to drink at the same time. If they have to wait and walk to water somewhere else as a herd or flock, much larger troughs are needed.

No matter what size troughs you use, make certain that your setup enables all animals, including submissive ones, to drink as much as they want. When animals have to leave a paddock and go to a single water source, submissive ones may never drink. To check on this you need to watch the herd or flock from a distance while they drink during various times on different days.

ELECTRIC FENCING

To manage grazing, livestock must be controlled. This means that you have to be able to place animals where you want them for as long as you want them there. Proper grazing management requires dependable fencing. Fortunately, having good fencing no longer is a problem.

Energizers

With the development of low-impedance electric fence energizers in New Zealand, controlling livestock became easier than it had been with ordinary fence chargers and nonelectric physical barrier fences. Energizers are effective because they produce an extremely short, high energy DC electrical pulse of about 6000 volts that charges fence wires once per second, or randomly every 1 to 7 seconds. The short pulse brings fence lines up to high voltage so quickly that the energy isn't easily drained off, even if rain-soaked plants touch the fence wires, or a short length of wire has fallen on the soil. Also, the short pulse generates very little heat, so there's almost no danger of an energized fence causing a fire.

Energizers are available to fit all needs and circumstances. There are models that run on 120 or 220 volts, 12-volt auto or marine batteries, solar energy, or flashlight or lantern batteries. The various models have different power outputs, suitable for different livestock and kinds of fencing. Discuss your fencing and livestock situation with an experienced dealer to select the energizer that will meet your needs. Usually it's best to buy a more powerful energizer than you think you'll ever need.

Energizers outperform all ordinary fence chargers in effective power by 2.5 to 20 times, depending on the size of the energizer. Although they dependably deliver a lot of power, energizers use very little electricity to produce it. For example, an energizer that will charge over 2 miles of five-strand fence or 60 miles of one-strand fence consumes only 8 watts of electricity!

Energizers cost somewhat more than ordinary chargers, but they're well worth the extra expense. Any extra costs are soon repaid in savings of time, labor, and materials needed to build and maintain an energized fencing system.

Because energizers don't short out easily, animals (livestock or predators) that touch the fence always receive a strong electrical shock that makes them respect the fence and avoid it. This dependable shocking power is why energized fences can be psychological or mental barriers, rather than physical barriers. Mental-barrier fences can be greatly simplified, and cost much less to build and/or maintain than physical-barrier fences, such as woven wire, multiple-strand barbwire or nonelectric smooth wire fences. (Never electrify barbwire, because animals or people can be cut badly by the muscle spasm resulting from the shock.) For example, a five-strand, spring-loaded high tensile fence for sheep, including the energizer, costs less than half as much per running foot than woven wire fencing.

Ordinary Chargers

An energizer makes it easier to control grazing, but it's not required, except for sheep. It's still possible to control cattle, horses, goats, and pigs with ordinary chargers. But because chargers are so easily shorted out by even one animal or wet plant touching the fence, they are not as dependable as energizers. An ordinary charger generally is one of the weakest links in a fencing system, and should be replaced with an energizer if you have any doubts about the charger's ability to control your livestock. Animals may not respect a fence powered by an ordinary charger, and can easily knock down all your paddock division wires if the charger shorts out for even a little while. A worse problem develops if animals learn how to go through poorly charged fences; they may be very difficult or impossible to retrain to stay in even with a well-charged fence.

If you have a good charger, its performance can be improved by:

• Making certain that it has a good earth return (well grounded). Follow the instructions on the next pages for making earth returns for energizers and chargers.

• Keeping all fence lines clear of tall grass and weeds, since ordinarily charged wires can't touch anything but insulators. If only one or two wires are used with cattle or horses, the animals can graze underneath the wire(s), keeping fence lines free of tall weeds and grass. Thistles and other unpalatable plants still will have to be cut, but the amount of time and effort needed for clearing fence lines will be less.

- Making certain that all insulators are in good condition, and that wires don't touch posts, trees, or brush.
- Cleaning fence lines of all old wire. An old barb wire fence running through weeds and brush and nailed to trees makes a good ground. If electric fence wires touch the barb wire, ordinary chargers and even energizers will short out.

Supposedly, sheep can be kept in with ordinary fence chargers, but not easily, because their wool insulates them from the electrical pulse produced by ordinary chargers. Keeping sheep in with an ordinary charger requires a very good charge, short fence, and few or no plants touching the wires. In my experience, once lambs or ewes learn how to go through a fence having low voltage, it's nearly impossible to keep them from doing it again, unless fence voltage is brought to at least 5000 volts. Ordinary chargers can't be used with the five-strand perimeter fencing needed to keep predators away from sheep or goats, because the lower wires must be close to the soil where a lot of plants can touch them. Net fences for small livestock also touch plants or soil, which would easily short-out an ordinary charger.

Earth Returns For Energizers And Ordinary Chargers
Energizers can fully charge very heavy fence loads and discourage even the most unruly livestock from going through fences. But to do this, they must have a good earth return. The performance of ordinary chargers also can be greatly improved by making certain that their earth return is adequate. Instructions for making a good earth return that usually accompany energizers are inadequate for many soils, especially those formed from glacial till.

An earth return that is too small or poorly constructed

not only limits the shocking power of fences, but electricity flowing in a fence may find other ways to return to the energizer/charger, resulting in stray voltage. Since the earth return is half of the electric fence circuit, I think it's worthwhile to explain one way of making a good one:

• Drive four to six grounding rods (galvanized pipes or copper rods) 40 to 50 feet apart, preferably into moist soil, beginning near the energizer or charger. Grounding rods should be 6- to 8-feet long, and should be driven straight down, if possible, to be in moist soil. If they can't be driven straight down, because of ledge or rocks, drive them in at an angle. Leave about 4 inches of rod sticking out above the soil surface, to make connections.

Rods can be placed in any formation, just so they're out of the way. If possible, place them so that one connection in the earth return can be made to a metal culvert under a road or lane, or to an unused well casing. The more powerful the energizer or charger, the larger the earth return needed. In dry sandy or glacial-till soils where grounding is difficult, 10 or more rods may be needed, if a metal culvert isn't available.

Don't use ground rods that are being used for anything else. Don't connect the earth return to anything above ground (e.g. woven-wire fence, metal building), because this could result in stray voltage.

• Clean connection points on the rods with a file, wire brush, or steel wool. Don't mix materials: use galvanized wire with galvanized pipe and copper wire with copper rods. Use insulated electric fence wire between the energizer and first ground rod to keep fence electricity separate from buildings and other electricity. After that use one length of bare wire to all grounding rods, to make certain that the connection is continuous. Double the wire for earth returns of larger energizers. Attach the wire securely to each rod with copper or stainless steel clamps or galvanized bolts and washers.

• Test the earth return. Before plugging the energizer or

charger into its power source, connect the unit to the fence and earth return. Then short-out the fence by laying about 200 feet of fence wire on the soil, at least 500 feet away from the energizer or charger, or attach the electric fence wire(s) to a metal object such as a metal post driven in the soil or a long, old woven-wire fence.

Then plug the unit in and test the earth return with a digital fence tester, by taking a reading between a grounding rod and soil about 4 feet away. You can also do this test by using your two hands spread well apart, touching the soil and the ground wire..... If the earth return is charged, improve the ground until it carries less than 300 volts when tested. If you hadn't connected to a culvert already, the easiest way to improve an earth return is to connect it with bare wire to the nearest metal culvert, even if it's several hundred feet away.

Lightning Arrestor And Choke (Induction Coil)
Energizers and chargers must be protected from lightning, which could induce high voltage on a fence line that would damage the unit and possibly start a fire in the building where the unit is kept. Lightning arrestors on fence lines provide an easy path for lightning-induced voltage to reach the earth. A choke (induction coil) installed at the energizer creates resistance that encourages lightning to follow a path to earth through arrestors, rather than through the energizer.

At least one lightning arrestor should be placed at the point where the hot wire from the energizer or charger attaches to the fence. For more protection, other arrestors should be installed at each corner of the fence, or one per mile of fence. For total protection during electrical storms, the energizer or charger should be unplugged, with the fence and ground terminals disconnected.

The ground for lightning arrestors can't be part of any other grounding system. Since lightning always takes the shortest and easiest path to the earth, the ground for lightning arrestors should be as good as the earth return for the energizer/charger. If the earth return for the energizer/charger is poor, arrestors won't protect the unit from damage.

Besides protecting your energizer/charger from lightning-induced high voltage, it's also a good idea to have an electrician install a large surge suppressor in your electrical supply. This will protect your unit from high voltage surges coming in from the power line.

Training Animals

To avoid discouraging problems, animals should be trained to respect an electric fence before turning them out to pasture. To do this, build either a strong multiple-strand electric fence or a wooden rail fence with one or two strands of electric wire toward the inside, around a barnyard or similar small area. An obvious barrier, such as wooden rails, on the other side of the electric wires trains animals to back away when they get shocked. Place some grain on the soil at various spots just inside of the electric wires, or attach strips of aluminum foil to wires to attract nose contact. Connect the energizer or charger to the electric training fence. Confine animals to the area for several hours (always have water available to them), until you're certain that at least some of them have been shocked and they all respect the fence.

FENCING MATERIALS AND CONSTRUCTION

We are fortunate that New Zealanders didn't give up on pasturing livestock as others did, but continued to develop fencing materials that would enable them to do a better job of managing grazing. Because of this and recent developments in the United States, New Zealand energizer and fencing technology has been combined with Yankee ingenuity to make available extremely effective electric fencing. These fencing materials are relatively inexpensive, easy to build and maintain, and long-lasting.

Let's look at some materials and construction techniques that are available. (Excellent detailed instructions on building high-tension electric fences and using portable fences are available from dealers of fencing materials, including those listed at the end of this chapter.)

High Tensile Wire
High tensile smooth fence wire is much stronger (1,000-pound breaking strength) than common soft smooth wire. This enables a high tensile fence to withstand a charging animal or winter snow loads without breaking. High tensile wire is elastic and flexible, but doesn't stretch like soft wire does. It has a heavier galvanized coating that makes it last at least three times longer than common soft galvanized smooth wire. (Special, heavily galvanized soft wire is now available.)

Springs And Tighteners
In fence runs of less than 450 feet, the elasticity of high tensile wires maintains tension on the fence. In fences over 450 feet long, a spring installed in each 1500 feet of wire maintains about 150 pounds of tension on the wire. In areas where there are snow and icing conditions, fence runs shorter than 450 feet should also have springs to prevent the wires from breaking. Tighteners placed in each wire near springs are used to take up slack in the wire and tighten it to the desired tension.

Posts
Corner posts must be strong, firmly set in the ground, and well-braced or anchored to hold high-tensioned wires. Corner posts should be at least 6 feet long, and can be 3- x 3-inch sawn hardwood, or 6-inch diameter softwood (usually cedar or treated pine). Trees located at corners can also be used with J-bolts and insulators to hold wires.

Line posts serve only to hold the wires up and maintain proper wire spacing. So line posts can be only 5 feet long, fairly light (1.5- to 2-inch diameter or square), and can be spaced up to about 150 feet apart. When line posts are widely spaced, lighter (1.25-inch diameter or square) battens that stand on top of the ground are used to maintain proper wire spacing. Battens and anchors are also used to hold the wires down where fences cross low spots in uneven terrain. Battens are needed about every 35 feet for five-strand fences, and every 50 feet for two- or three- strand fences. Four-foot, 1.25-inch diameter or square line posts are used instead of battens for keeping one-strand fences in place.

Using self-insulating posts decreases construction and maintenance costs because no insulators are needed in attaching wires to these kinds of posts. Hardwood line posts with sharpened points can be easily driven into the ground with a post driver or light sledge hammer. Slotted line posts and battens hold wires at proper spacings. Wires are held loosely in the slots with heavy wire clips, or can be attached with staples to unslotted posts. This allows fence wires to move in the slots, and keeps the fence flexible so that it can withstand stress, and spring back into place without breaking.

Insulators

Pay particular attention to insulators, because poor insulation of charged wires can result in needing a higher powered energizer than otherwise would be necessary. Insulators must be used wherever wires are attached to anything but self-insulating posts. Use the best insulators available. Don't use the single-nail kinds, because they don't insulate well and aren't strong enough. Insulators must be strong and large enough to support long spans of high-tension wire, and allow wires to slide freely.

Wire Spacing

Fence wire spacing can be as follows for:

Dairy, Beef, Horses:

High-tensile perimeter fence: one strand at about 33 inches; two strands at 20 and 36 inches; or three strands at 16, 28, and 40 inches.

Subdivision fences (can be high-tensile wire, soft smooth wire, polywire, or polytape):

• One strand for cows, yearling heifers, and horses at 28 to 33 inches from the soil surface (46-inch high fences are needed for some horses).

• Two strands for cows with calves, 6- to 12-month-old heifers, and horses with foals: 20 and 36 inches from the soil surface.

Sheep And Goats:

Perimeter fence: five-strands of high-tensile wire for keeping out predators at 4-6, 10-12, 16-18, 24-26, and 36 inches from the soil surface. An additional wire may be needed at 48 to 60 inches for goats that jump.

Internal subdivisions: 3 strands of high-tensile wire, soft smooth wire, or polywire at 10, 18, and 30-36 inches.

Pigs:

Two strands at 7 and 17 inches, or

Three strands at 6, 13, and 23 inches.

Cutout Switches

Paddock layouts should be designed so that separate sections can be shut off with DC cutout switches. Convenient switch locations are at gates, corners, and places where one large pasture area ends and another begins. The switches allow you to shut off unused sections of paddocks and put full electrical power into the section(s) containing animals. In case of a problem, having paddocks in separate sections allows you to test each section in sequence, starting with the one nearest the energizer or charger. If predators are a problem, always keep all perimeter fences charged.

Portable Fencing

Lightweight portable fencing is easily moved and reused at many locations during a grazing season. It costs more per foot than permanent fencing, but much less is needed. Since it takes the place of large amounts of permanent fence, portable fencing can save a lot of money. Portable fencing can be used with energizers and solid-state ordinary chargers, but not with "weed-burners". The long pulse of "weed-burners" creates heat in fence wires, which would melt plastic portable fencing.

Polywire

Polywire is a polyethylene twine braided together with six or more strands of overlapping stainless steel to carry the electric charge. It comes in rolls of various lengths. Polywire has many uses, including subdividing large permanently fenced paddocks, temporarily fencing around irregular areas or hill contours, making gates, and looping down over stream beds to keep large animals from walking beneath permanent paddock division wires. When used to subdivide large paddocks, polywire is supported on plastic or fiberglass posts, or on tumblewheels. Polywire shouldn't be used as a mainline fence to feed power to other fences, because it doesn't conduct electricity well enough.

Polytape

Polytape is a polyethylene tape 0.5 to 1.5 inch wide that has 5 to 11 strands of stainless steel embedded in it. Polytape comes in 330- to 1600-foot rolls, and is used for the same purposes as polywire, but is more easily seen by livestock than polywire. This may be especially helpful with some animals that can't see small-diameter wire very well, or that are especially afraid of electric fence, such as horses. It's very useful as a training fence.

Reels
Completely insulated hand reels are used to roll up polywire or polytape. The reels are light, well balanced for handling ease, and can be hooked onto perimeter fences. (Inexpensive plastic electric cord reels available at hardware stores are useful for polywire.) It's possible to attach three to five reels to a short steel post, that then allows three to five polywires or tapes to be positioned at the same time for internal fencing of small livestock. It's also possible to have three shorter separate strands of polywire on one reel, and position the three polywires at the same time.

Rolling Fence Wheels
Rolling fence wheels (e.g. Gallagher tumblewheels) essentially are six-spoked, rimless rolling fence posts, that hold up and are held up by polywire or polytape ahead of and behind grazing animals. They're quickly and easily moved, inexpensive, and made of very light noncorrosive materials.

Electric Net Fencing
Electric net fencing for small livestock consists of six or seven horizontal strands of polywire woven to form a large-mesh net, with vertical braided or solid strands of nonconducting polyethylene. About every 12.5 feet a round plastic post is incorporated into the netting to hold the fence up.

Plastic And Fiberglass Posts
Plastic posts with wire holders molded into the posts at proper wire-spacing intervals are available to make installing multiple-strand portable fencing very easy. They also have a large pointed rod on the bottom that goes into the soil, and a lip to step on for pushing them in. This means that you can carry a bundle of 20 or more posts, pick one from the bundle, and step it into the soil without

setting the bundle of posts down.

Round fiberglass posts with adjustable plastic or metal wire holders are useful for making single-strand polywire or metal wire fences. Round and T-shaped fiberglass posts are available, but the round ones are better made and easier to use. The T-shaped posts give off a lot of small splinters that are very painful and nearly impossible to find and remove. Round fiberglass posts also give off some splinters, so it's a good idea to wear gloves when handling any fiberglass posts. Also, wires must be held to the T-shape posts with metal clips that don't allow wires to move, so they can't be tightened without removing all of the clips. Another problem is that every time an animal hits the fence hard enough, a lot of clips pop off the T-posts; these could end up inside your animals.

Physical-Barrier Fencing
If you already have good physical-barrier fencing, by all means use it. I don't want you to get the impression from this discussion that to control grazing you have to spend a lot of money building new fences. Quite the contrary: spend as little as possible getting started. This book is to help you manage your pastures in a way that will make your farm more profitable. I certainly don't want you to go further into debt in setting up this system. Because electric fencing costs less than physical-barrier fencing, it's usually best to replace physical-barrier fence as it wears out with long-lasting, low-maintenance high tension electric fence.

Physical-barrier fencing works perfectly well in most situations, especially as perimeter fencing for large livestock. Just make certain that the fencing is tight and in good condition. When used for paddock divisions, however, barbwire and nonelectric smooth wire don't work very well, because animals gradually loosen it up by reaching underneath and through it to eat forage in adjacent paddocks. It's best to use portable or low tension electric fencing to subdivide areas within good physical-barrier perimeter fencing.

The useful life of physical-barrier fencing can be extended by attaching special offset brackets with insulators to the fences, and adding an electrified smooth wire to them. This prevents animals from pushing against and loosening the fences. This offset electrified wire can also be used to carry power to portable or permanent subdivision fences. Don't attempt to run an electrified strand along a nonelectric fence that is in poor condition: it will cause you more problems than it's worth, by shorting out your electric fence system.

In some countries neither energizers nor ordinary chargers are available, or they're too expensive for farmers to buy because of import taxes. Although electric fencing makes it easier to subdivide pastures, physical-barrier fencing certainly can be used to build paddocks. Rock walls, split rails, stumps and brush, bamboo, living briars and hedges, woven wire, and multiple-strand barbwire or smooth wire are examples of materials useful for making physical-barrier fences. It doesn't matter how pastures are subdivided, just so they're divided enough to provide adequate recovery periods for the plants under local conditions, and short occupation periods.

SOURCES OF ENERGIZERS & FENCING MATERIALS

Your local Extension Service and Natural Resource Conservation Service should have information about dealers that sell energizers and fencing materials, and build high tension electric fencing systems. Check telephone book yellow pages. If the information isn't available, this list of companies should help you to find a nearby dealer:

Dailey Fence & Supply, 5385 Edgemoor Road, Adamsville, OH 43802. 614-796-6531.

Gallagher Power Fence, Inc., P.O. Box 708900, San Antonio, TX 78270-8900. 800-531-5908. Ask for excellent free fencing guide.

Grassland Supply, R 3 Box 6, Council Grove, KS 66846. 800-527-5487.

Jeffers, P.O. Box 948, West Plains, MO 65775 and P.O. Box 100, Dothan, AL 36302. 800-533-3377.

Kencove, 111 Kendall Lane, Blairsville, PA 15717. 412-459-8991.

Kentucky Graziers Supply, 1929 S. Main Street, Paris, KY 40361. 800-729-0592.

Kiwi Fence Systems, Inc., RD 2 Box 51A, Waynesburg, PA 15370. 412-627-8158.

McBee Agri Supply, Inc., Rt. 1 Box 121, Sturgeon, MO 65284. 314-696-2517.

Parker McCrory Manufacturing Co. (Parmak), 2000 Forest Ave., Kansas City, MO 64108.

Perfect Pastures, 2339 CTH I, Highland, WI 53543. 800-829-5459.

Premier Fence Systems, P O Box 89, Washington, IA 52353. 800-282-6631. Ask for excellent free fencing guide.

Tipper Tie, P O Box 866, Lufkin Rd, Apex, NC 27502. 800-441-3362.

Trident Enterprises, Inc., 9735 Bethel Road, Frederick, MD 21701. 301-694-6072.

Twin Mountain Supply, P.O. Box 2240, San Angelo, TX 76902. 800-331-0044 (800-527-0990 in Texas).

West Virginia Fence Corporation, U.S. Hwy 219, Lindside, WV 24951. 800-356-5458.

REFERENCES

Cadwallader, T.K. 1996. Electric fencing basics: constructing an earth return system. *Stockman Grass Farmer*, 53(5):13-15.

Daigle, P. 1996. Lanes that keep dairy animals high and dry. *Stockman Grass Farmer*, 53(7):11-13.

Davis, L., M. Brittingham, L. Garber, & D. Rourke. 1991. *Stream Bank Fencing.* Publications Dist. Ctr., Penn State Univ., University Park, PA 16802. 11 p.

Gallagher Electronics Ltd. 1985. *Gallagher Insultimber Power Fencing Manual.* Hamilton, New Zealand. Brochure. 24 p.

Gallagher Electronics Ltd. 1985. *Gallagher Temporary Power Fencing Systems.* Hamilton, New Zealand. Brochure. 6 p.

Gallagher Electronics Ltd. 1985. *Tumblewheel: the Simple and Effective Time Saving Aid to Strip Grazing.* Hamilton, New Zealand.

Green, J. 1993. Grazing habits and forages for goats. *Stockman Grass Farmer.* 50(6):8.

Jones, V. 1993. Sundry grazing management advice. *Stockman Grass Farmer.* 50(11):8-9.

Martyn, R. 1994. Dairy farm water supply basics. *Stockman Grass Farmer.* 51(4):13,30.

Nation, A. 1992. What's New... Grazier's Gear. *Stockman Grass Farmer.* 49(12):14.

Patenaude, D. and J. Patenaude. 1991. *The Grazing News.* Perfect Pastures, Route 1, Highland WI. Newsletter. March.

Pel Electric Fence Systems. *Instruction Manual.* 28 p.

Premier Fence Systems. *A Guide to Fencing that Works!.* Box 89, Washington, Iowa. 51 p.

Quillin, R. 1992. Portable livestock watering system easy as child's play. *Stockman Grass Farmer.* 49(10):17.

Salatin, J. 1996. Let your land guide you to paddock layout. *Stockman Grass Farmer.* 53(1):33, 39.

Swayze, H.S. 1984. *Gallagher Spring-Tight Power Fence (Construction, Materials, and Costs).* Tunbridge, Vermont. 8 p.

7
Livestock Production

Miraculous power and marvelous activity --
Drawing water and hewing wood!
Pan-yun

Animal evolution has been greatly influenced by diet available, so animal and plant development have been closely linked. Since plants contain cellulose, which is a polymer of glucose that makes up part of the fiber in plants, it's interesting that animals didn't evolve enzymes capable of breaking down cellulose. It's especially surprising because cellulose is one of the most abundant naturally occurring organic compounds that can be used for food.

The grazing animal's life style may have had a lot to do with how things turned out. The defenseless grazers probably would leave the forest cover only to fill up quickly on plants growing in clearings near the forest edge. Once full, they would return to hiding places to digest the food. Instead of cellulase enzymes, grazing animals developed digestive tracts in which vast numbers of symbiotic microorganisms live and digest by fermentation the fibrous and soluble plant constituents. These microorganisms effectively became the animal's cellulase-producing tissue.

This kind of digestive symbiotic system reached its fullest development in ruminants (e.g. cattle, sheep, goats) which contain a large fermentation vat called the rumen. Ruminants use pasture forage by first swallowing it into the rumen for fermentation. Large coarse forage material (cud) is regurgitated, rechewed, and swallowed into the rumen again. Most of the soluble nutrients are consumed rapidly by rumen microorganisms.

Some unfermented fiber does leave the rumen, but it and other food material and microorganisms that pass out of the rumen are worked on by normal enzymatic digestion in the lower gastrointestinal tract. The rumen absorbs nutrients produced by microbial fermentation, such as acetic, propionic, and butyric acids, which contribute 60 to 75% of the animal's energy requirements. Rumen microorganisms also produce amino acids that are essential to ruminants for making proteins. Vitamin K and the B-complex vitamins are all synthesized by microorganisms in the rumen.

Nonruminants (e.g. horses, pigs) have much smaller digestive tracts than ruminants, and feed consequently passes through them faster. Feed passes through a horse's stomach in about 24 hours, whereas feed remains in a cow's rumen for about 72 hours. Compared to ruminants, relatively little microbial action occurs in nonruminant stomachs, and what does occur is located in the cecum, after the small intestine. This contrasts with the rumen, which is located before the small intestine. For this reason, ruminants have an advantage of greater absorptive capacity compared to nonruminants.

Feed digestion in poultry (e.g. chickens, turkeys, ducks, geese) occurs in two organs: the proventriculus and the gizzard. Hydrochloric acid and enzymes are secreted into feed by the proventriculus, before the feed passes to the gizzard where it's ground and mixed with digestive fluids.

What all this means simply is that nonruminants and poultry have different nutritional requirements and must eat more frequently than ruminants. Despite differences, most or all of the nutritional needs of all livestock can be met on properly managed pasture.

To estimate potential livestock productivity from pasture, we must know three things:

• Nutritive value and yield of the pasture forage.
• Feed requirements of grazing animals for a particular level of performance (e.g. rate of weight gain or milk production) within the physiological states of growth, fattening, lactation, or maintenance.

• Total animal need and distribution of need for forage during the grazing season.

WATER

Water is the nutrient needed in largest amounts by all livestock, and is the nutrient that first limits production. Clean, cool, good-quality water must be available at all times to grazing livestock (see chapter 6). Water needs vary depending on size of animal, milk production, activity, dry matter intake, rate of gain, pregnancy, air temperature, and relative humidity.

Always consider water quality and availability when planning and setting up for any kind of livestock production, and when evaluating production problems. Even though your water supply passes farm inspection requirements, it doesn't necessarily mean that it's suitable for livestock to drink. Animals are sensitive to temperature, taste, odor, and pH of water. Bacteria and algae in water can harm cows without affecting milk quality. Stray voltage can limit or prevent animals from drinking. Water quality problems can affect livestock without affecting humans drinking from the same source.

Just because a water source has always provided good quality water doesn't mean that it always will. Changes in the water table level or amounts of water used in surrounding areas can influence water quality. Water quality of springs especially responds quickly to such changes.

Water quality and availability should be monitored annually to make certain that no problems exist. At the very least water should be tested for bacterial contamination. Follow sample collection instructions of the laboratory that will analyze the water samples. It's best to collect multiple samples (2 days apart) of the water where livestock drink. Keep the samples refrigerated, and have them analyzed as soon as possible.

Clean water tanks or tubs regularly so that the water

doesn't carry dirt or pathogenic bacteria. A dirty water tank easily spreads parasites and diseases to livestock drinking from it. Algae that grows in tanks that haven't been cleaned is toxic and causes scours in cattle. If you wouldn't drink from a water source, don't make your animals drink from it!

All animals drink less water when grazing lush pasture forage that contains 75 to 80% water. For example, sheep may go for days without drinking when they're grazing lush pasture. But even if animals don't drink, keep water available to them. On hot days check to make certain that the water is cool enough to drink. You may be surprised to find that the water is too hot to touch, let alone drink!

Daily water needs (including water in forage) in gallons per animal per day are: 12 to 36 for milking cows, 5 to 20 for full-grown beef cattle, 1 for mature sheep, 0.5 for weaned lambs, 5 to 6 for milking goats, 10 to 12 for horses, and 0.5 to 1.5 for pigs. Poultry also must have free access to fresh, clean water at all times. Of course, ducks and geese like water for bathing and playing in, besides drinking.

PASTURE NUTRITIVE VALUE AND LIVESTOCK MINERAL REQUIREMENTS

Protein, energy, and fiber contents of pasture forage determined by laboratory analyses are indicators of its nutritive value. High-quality forage contains low amounts of fiber and high levels of protein and energy; in this respect legumes at the same stage of growth are better than grasses. Season and stage of growth within a season affect the nutritive value of pasture forage. Protein, energy, and fiber values of pasture plants are most conducive to animal production when the plants are grazed at young stages of growth.

In properly managed pastures, crude protein content of the forage can average more than 22% over a 6-month grazing season. Based on National Research Council

(NRC) standards, this qualifies forage from well- managed pastures to be defined as a protein supplement! (Protein supplements are feedstuffs which contain more than 20% crude protein.) Forage from excellent pastures can contain as much or more energy as in barley, oats, triticale, or wheat grain. This is enough to meet most or all of the energy needs for all but the highest producing animals, depending on intake. Dry matter (DM) content of forage of well-managed pastures ranges between 20 and 25%.

Grazing management indirectly affects pasture nutritive value by controlling the plants' stage of growth. Short (3-4") leafy pasture is typical of good feed for sheep or goats when they are turned into a paddock. Longer (5-6") leafy plants characterize pasture for cattle or horses when it's ready to be grazed. Preflowering and flowering stages correspond to forage ready for haying or ensiling.

When grasses grow past the head-emergence stage, digestibility decreases. Periodic grazing, however, returns pasture plants to physiologically young stages of growth. Because of the effects of temperature and daylength on plant growth, the leaf-to-stem ratio, protein, energy, and fiber contents still change in plants that have been grazed. But the decrease in digestibility and increase in fiber of grazed plants are very much less than in plants allowed to grow uncut or ungrazed.

Besides protein, energy, and fiber, pasture forage also contains low amounts of fat, and variable amounts of minerals and vitamins, depending on soil fertility and plant growth stage. Mineral needs of livestock vary widely. The best ways to make certain that adequate levels of minerals are present in pasture forage is to feed the soil by applying manure, lime, and fertilizer, and maintain as complex a plant mixture in the pasture as possible. No one species of plant is good at everything, and one plant species can't satisfy all of the mineral and other nutritional needs of grazing livestock. Green forage from well-managed pastures that have mixed plant communities usually contains all of the vitamins needed by livestock.

You may improve your livestock's health and production by providing soluble mineral mixes through drinking water or as loose granular material in paddocks. In New Zealand research, animals supplemented with soluble mineral mixes ate less forage, grew faster, and had fewer health, stress, and parasite problems.

Salt and Trace Elements
Sodium and chlorine are needed by all animals, especially plant eaters. Since forage plants may not contain enough of these elements to satisfy animal needs, salt (sodium chloride) should be available at all times to animals on pasture. If animals bawl to get out of a paddock right after entering it, there might be a mineral imbalance that makes the forage unpalatable. Low sodium content makes forage unpalatable. New Zealand farmers fertilize with salt to make forage more palatable. Salt should be provided free-choice as loose granules rather than as blocks, because animals may not get enough salt from blocks to meet their needs, especially during hot dry weather.

If there's any doubt about the availability of trace elements such as cobalt, copper, iodine, manganese, selenium, or zinc in your pasture soils, provide a mineral mix or salt that contains the needed trace elements.

Soils in northeastern North America generally contain low amounts of selenium. Cows deficient in selenium, from grazing forage grown on low-selenium soils, scour as they walk along, leaving strung out manure. They hold their tails high to keep their tails and rears clean. Other symptoms of selenium deficiency in adult livestock are weak muscles, arched or hunched backs, high levels of mastitis and somatic cell counts, low solids percentage in milk, low milk production, miscarriage, foot and leg problems, weak immune system, retained placenta, delayed cycling, poor conception rates, and heart attacks.

Selenium level in the total ration should be at least 0.3 ppm, with at least 0.2 ppm in pasture forage. Plants accumulate different amounts of selenium, so the amount in pasture forage varies not only because of soil parent material, but with the plants growing on the soil. For example, orchardgrass contains relatively high levels of selenium, but white clover generally has a low content. Raising soil pH by liming helps to make selenium more available. Including an adequate amount of vitamin E in the ration helps animals absorb selenium.

Animal response to selenium supplementation can be quite marked. For example, milking cows grazing selenium-deficient pasture in New Zealand increased milk production 21% within 7 days of beginning to receive 2 ml/cow/day of a 5% selenium solution! Before supplementing with selenium, check its level in animals and total ration to make certain that it's needed and how much you should provide.

Providing trace elements is especially critical if you aren't feeding grain or concentrate supplements, which generally contain added trace elements. In some situations the greatest response to supplemental feeding may be due to minerals or trace elements contained in supplements, rather than to their protein and/or energy. It's usually a lot less expensive to provide needed minerals and trace elements with or without salt, rather than in grain or concentrates.

Be sure to provide trace element mixtures suitable to the kind of livestock you're raising and your local conditions. Ask other farmers, your local Extension agent and/or feed sales person for information about salt, trace element, and other mineral needs for your area and livestock.

Carefully place the salt and other mineral supply at a location in each paddock where you want the animals to spend most of their time in that rotation. By placing salt and minerals at different locations in each paddock in every rotation, you can distribute the animals and their manure more uniformly throughout the paddocks, and avoid overgrazing areas around the salt and minerals. Placing the salt and mineral supply in the middle of an area containing unpalatable weeds will help to destroy the weeds by trampling.

The amounts of salt needed (lb/animal/month) are: 3 to 5 for mature cattle, 1 for mature ewes, 1.5 for milking goats, and 5 pounds for horses. The salt requirement of horses increases greatly when they work hard and sweat a lot. Salt is also one of the minerals needed in largest amounts by pigs, and should be available to them free-choice at all times. Poultry require salt at the rate of 0.2 to 0.5% of their diet, which they can obtain if loose salt is made available. Lactating animals need more salt than dry ones. Young animals need less salt than mature ones.

Calcium And Phosphorus
Next to sodium and chlorine, grazing animals are more likely to experience a deficiency of phosphorus than any other element. Although most forage plants contain adequate amounts of calcium, they may be low in phosphorus, especially if the phosphorus level in the soil is low. Legumes are excellent sources of calcium, but grasses contain less calcium. Forage analyses can easily tell you how much calcium and phosphorus your pasture forage contains. Feeding the soil by applying rock phosphate or phosphorus fertilizers according to soil-test recommendations, usually increases forage phosphorus level and yield of pasture plants. If it's not possible to apply all the phosphorus that's needed because of the expense involved, or if phosphorus fertilization fails to raise phosphorus content of the plants to adequate levels, phosphorus supplements should be made available free-choice to the animals.

Calcium and phosphorus use by grazing animals

depends on:
• An adequate supply of available forms of both elements.
• A suitable ratio between them.
• Adequate Vitamin D from sunlight or the ration to enable calcium and phosphorus to be assimilated and used. If there's plenty of Vitamin D available, the ratio of calcium to phosphorus is less important. Conversely, if the ratio of calcium to phosphorus is satisfactory, less Vitamin D is needed.

Calcium to phosphorus ratios in the total ration for ruminants should be in the range of 2:1 to 4:1, except for dry cows, which need a ratio of 1.2:1 to 1.7:1. Usually the calcium to phosphorus ratios for horses should be about 1.1:1. Growing pigs and poultry require a ratio of 1.2:1 to 1.5:1. It's always important to have more calcium in feed than phosphorus, but not too much more. As production increases, needs of calcium and phosphorus (and all other minerals) also increase.

If calcium or phosphorus supplements are needed, the amount and source used depends on the mineral composition of the total ration. If only calcium needs to be supplemented, use ground limestone or oyster shells. If only phosphorus is needed, use monosodium phosphate, disodium phosphate, sodium tripolyphosphate, or feed-grade phosphoric acid. If both calcium and phosphorus are needed, use dicalcium phosphate, steamed bone meal, defluorinated rock phosphate, or a commercial mineral mix. Provide minerals free-choice, mixed with about 40% salt to ensure that adequate amounts are eaten. Crushed oyster shells should always be provided for grazing poultry to satisfy any need for extra calcium.

Magnesium
The most common problem of magnesium deficiency in grazing animals is called grass tetany, grass staggers, or hypomagnesaemia. Cattle are more susceptible to grass tetany than sheep or goats, and females are most

susceptible, especially when they're milking or pregnant.

Grass tetany occurs most frequently in animals grazing lush grass-dominated leys, highly fertilized with nitrogen and potassium. If soil pH level is low, plant uptake of magnesium is reduced even more. Grass tetany is rare or nonexistent in animals grazing pastures growing on soils that developed from dolomite. It occurs more commonly in the spring among unsheltered animals during cold stormy weather, especially within 2 weeks to 2 months after giving birth. Animals grazing pastures that contain legumes and other broadleaf plants (e.g. dandelion, plantain, chicory), are less likely to develop grass tetany, because broadleaf plants contain more magnesium than grasses.

Grass tetany results from a low level of magnesium in the blood, which may be due to eating forage containing less than 0.2% magnesium. An imbalance of ions occurring when the ratio of potassium and sodium to calcium and magnesium is unfavorable in forage may indirectly result in low magnesium intake and absorption.

To prevent grass tetany, pasture fertilization should be balanced by applying magnesium along with other nutrients. Manage pastures so that clover growth is encouraged. Also, animals should be given a magnesium supplement, especially during the spring in areas where grass tetany is a problem.

Animals affected with grass tetany become nervous, and develop leg muscle spasms and convulsions. Animals usually recover if treated with a timely injection of magnesium gluconate and sedatives.

Protein
The amount of protein needed, if any, in a concentrate supplement depends on the quality and kind of forage that animals are eating in a pasture, and on the kind of animal, its physiological state, and its production level. In general, as protein content of forage increases, less protein

supplement is needed. The total ration (forage plus protein and energy supplements) should contain at least 14% total protein, except for ewes and beef cows, which don't need as much. High-producing dairy cows may need about 17% total protein to digest the energy in the extra feed they require, and to make milk.

Under good feeding conditions of intensively managed pastures on fertile soils, grazing animals usually consume more protein than they need. For example, pastures under management intensive grazing in Vermont have averaged more than 22% crude protein in the forage over a 6-month grazing season -- more than enough for most livestock, if pasture forage dry matter intake is adequate. If protein needs can be met with pasture forage, a lot of money can be saved in feeding costs, because protein supplements usually are very expensive.

Energy
Since protein deficiencies usually don't exist for most animals (except possibly for high-producing dairy cows) grazing well-managed pastures, such pastures may be evaluated only in terms of the amount of energy in the feed that's available to the animals after digestion and metabolism. In terms of quantity, energy is the most important part of an animal's diet. All feeding standards and rations are based on some measure of energy, with additional inputs of protein, amino acids, essential fatty acids, vitamins, and minerals.

When there's plenty of forage to graze (adequate forage allowance), the amount of energy in the forage that's available to animals is related to the animals' intake of the forage. Intake is greatest for pasture forage that's low in fiber and high in protein and energy. Such forage is very palatable to grazing animals. As pasture forage quality decreases, the bulk of the forage eaten limits intake, because the larger amount of fiber in the plants breaks down more slowly and keeps the gastrointestinal tract full of fiber longer. (See chapter 8 on Feed Planning for further discussion about energy in livestock production.)

PASTURE EVALUATION

Sampling
To evaluate pasture forage nutritive or feeding value and dry matter intake, representative samples of the forage must be taken and analyzed. Unfortunately, this is easier said than done. Entire books exist on techniques of sampling pasture forage, and none of the methods is completely satisfactory. The problem is that the analytical results are only as good as the sample. The difficulty lies in taking samples that represent the general nutritional state of forage present in the pasture, and the forage that animals actually graze.

Forage Feeding Value
We use a pasture forage sampling method for nutrient analysis that's simple, inexpensive, and gives good results. It involves walking in a zig-zag path across a paddock just before it is grazed, and taking 30 to 40 subsamples of the forage. For each subsample take the amount of forage that you can hold and pull off with your thumb and two fingers, taking only what animals are likely to graze. Place the subsamples in a plastic bag as you collect them, combining them all into one composite sample that represents the overall forage in the paddock.

Sample only from areas that will be grazed, not in rejected forage spots around manure. Rejected forage around manure tends to be higher in phosphorus and potassium, and grows about twice as fast as forage of grazed areas. Rejected forage also becomes more mature and decreases in feeding value as the season progresses. Since the forage is rejected, there's not much point in sampling and analyzing it.

Label the samples and freeze them until you can get

them analyzed. Have samples analyzed as soon as possible after taking them, so that you know the quality of forage that your animals are eating and can adjust your in-barn feeding program accordingly. Using overnight shipping and FAX services, it's possible to have analysis results within 2 days of sampling. You can get a good estimate of the quality of forage available by sampling one or two representative paddocks every 2 to 3 weeks.

Forage Dry Matter Intake
Variation in pasture forage intake by grazing livestock has a major influence on animal performance. Sward botanical composition, leaf-to-stem ratio, pre- and postgrazing pasture mass, and forage allowance probably have more important effects on forage intake than nutritional aspects. An objective of grazing management is to influence these physical aspects to make available easily grazed forage, so intake can be as high as possible.

Estimates of dry matter intake of grazing livestock can be made for individuals using direct or indirect animal techniques, or for groups by pasture sampling. All estimates are just that: they aren't precise, no matter who made them or where they came from. All techniques used to make estimates are subject to error due to variation in pasture swards, animals, and sampling techniques. Use estimates simply as guides to help you in managing grazing.

The most common animal technique involves an indirect estimate of intake from laboratory determinations of pasture forage digestibility, and indirect or direct measurements of poop output of grazing animals. By this method: DM intake = poop output ÷ (1-digestibility)

Average dry matter intake for groups of livestock is estimated from the difference between pre- and postgrazing pasture masses in a paddock of known area. This method is best used over short occupation periods of less than 4 to 5 days. Intakes estimated by pasture sampling generally are 30 to 40% lower than intakes estimated by animal techniques. By this method:

$$\text{DM intake} = \frac{\text{pregrazing mass - postgrazing mass}}{\text{number of animals grazing x days}}$$

The values for intake (and protein and energy requirements) in the tables at the end of chapter 8 are based on New Zealand estimates, which were obtained from animals grazing pasture. This contrasts with values available in the United States that were obtained with stall-fed animals. A lot of assumptions have to be made when using stall feeding to get information about grazing animals; not all assumptions hold true in real life.

Another way to evaluate forage quality and intake is to look at animals' poop. Dung consistency reflects forage quality and rates of dry matter intake and passage. Production output depends on feed intake and digestion speed. The faster forage passes through an animal, the more it can eat, and the more milk it can produce or weight it can gain. Runny dung indicates that an animal mainly is eating highly digestible leaf, with a high rate of passage and production potential. Dry dung indicates intake of fibrous, poor-quality material with a low passage rate and production potential.

MILKING COWS
More than anything else, feed determines the profitability and productivity of dairy cows. Feeding expenses account for 45 to 65% of the total cost of producing milk. The lower these costs are, the more profitable the dairy farm will be. Pasturing is the least-cost way of feeding dairy animals and other livestock.

Supplements
The process of managing rations for dairy cows is the same whether they're fed on pasture or in confinement.

In both situations we need to estimate forage quality and requirements, and supplement accordingly, if it's profitable to supplement. It's somewhat difficult to estimate forage quality and dry matter intake from pasture, because of variation in the swards that cows graze. But when cows are fed in loose-housing, dry matter intake per cow isn't known for certain either.

Depending on pasture forage composition (grass, legume, forb), availability, palatability, protein and energy content, and cow liveweight, certain amounts of milk and butterfat can be produced without energy or protein supplements. Legume content of pasture forage is extremely important.

For example, in Ohio research, unsupplemented (except for minerals and vitamins) cows grazing alfalfa produced 45 lb milk/cow/day. Unsupplemented cows grazing high-quality orchardgrass without any legume produced only 30 lb milk/cow/day. Cows grazing orchardgrass and fed 15 to 20 lb of energy and protein supplement per cow per day produced 45 lb milk/cow/day.

Research in Ireland, that included unsupplemented animals in experiments, showed that pasture-based dairy farmers should question supplement recommendations developed for cows fed stored forages. Energy and protein supplementation needed with hay and silage feeding was found to be unnecessary when animals grazed green pasture. As long as sward height was suitable for high intake (i.e. 6 inches pregrazing, 3 inches postgrazing after producing animals) and quality was adequate (no seedheads or stems, and contained 30-50% white clover), there was no economic benefit from supplementing cows. Although milk yield increased with supplementation, costs increased at the same rate.

Irish researchers also found that:
• Concentrate and/or grain supplementation of pastured dairy cows was profitable only during periods of pasture forage shortages. These results, of course, reflect Irish

milk and concentrate and grain prices, their pasture forage quality, and the forage-converting efficiency of their cows (Fresians and Jerseys).

• Feeding high-starch grains (e.g. corn, barley) worsened spring milk fat depression, and lowered pasture forage digestibility by increasing rumen acidity; this decreased forage intake. Feeding sodium bicarbonate didn't affect milk yield or fat percentage.

• Feeding high fiber supplements (e.g. oats, cottonseed, beet pulp) didn't decrease milk fat content, and weren't profitable to feed if pasture forage amounts and quality were adequate.

• All supplements substituted expensive feed for inexpensive pasture forage. If pasture forage quality and quantity were adequate, each pound of supplement eaten substituted for one pound of pasture dry matter. Cows decreased grazing time 10 minutes/cow/day per pound of supplement dry matter fed.

• Recommendations for feeding supplemental bypass protein to cows grazing well-managed pasture appear to be invalid.

• Feeding hay and/or grass silage to cows grazing well-managed pasture decreased total grazing time 20 minutes/cow/day per pound of supplement eaten, and reduced milk and protein yield.

• Feeding corn silage to cows grazing well-managed pasture decreased milk yield.

• Supplementing growing replacement heifers grazing well-managed pasture was unprofitable.

New Zealand dairy farmers also feed little or no grain or concentrate supplements. Their milk sells for about 6 cents/lb and grain or concentrate costs about 10 cents/lb, and there's no economic benefit from supplementing.

In the USA milk sells for about 12 cents/lb and corn costs 4 cents/lb, so we have a different situation from that of Ireland and New Zealand. Larry Tranel, a University of Wisconsin Extension Agent specializing in farm financial management, has calculated how it could be profitable to supplement cows in the USA:

- Each pound of grain fed to dairy cows results in about a 0.25- to 1.5-lb increase in milk yield, depending on stage of lactation, and displaces about 2/3 lb of forage eaten. If pasture forage costs 1.75 cents/lb ($30/ton) and corn costs 4 cents/lb, it actually costs 2.8 cents/lb of grain fed [4 cents - (2/3 x 1.75 cents) = 2.8 cents]. So there needs to be at least a 1/4-lb yield response (worth 3 cents) to break even. If there's a 1-lb milk yield increase per pound of grain fed, it is a 300% return on money spent for grain.

While this milk price/grain cost relationship holds, it may be profitable to supplement dairy cows on well-managed pasture here, up to a certain point. Research is needed to find that point, and should take into consideration all aspects of cost (e.g. labor, machinery needed, soil eroded and pollution caused during grain production). This price/cost relationship probably will change when the recent international trade agreements are implemented. What will happen if the price that American farmers receive for milk drops to the international price level?

Energy

Cows producing the most milk require excellent quality pasture forage, to achieve the needed intake of energy for high production levels without energy supplements. Smaller cows can't achieve as high a level of production on poor forage as larger cows can. This probably is due to the physical limitation of how much forage can be eaten in a day by different size animals. But smaller cows produce more milk per 1,000 pounds of body weight than larger cows, and are therefore more efficient than larger cows. Smaller cows (e.g. Jerseys) also make up for producing a lower volume of milk than larger cows (e.g. Holsteins) by putting more butterfat in their milk and eating less total feed.

Another important point to note here for milking cows and all grazing animals, is that at a particular level of productivity, a greater intake of energy is needed when animals graze low-quality pasture, compared to that needed when they graze high-quality pasture. This is because they walk around less searching for preferred feed, and expend less energy both in obtaining the feed and in digesting it when they graze high-quality pasture.

Providing energy supplements can be beneficial, up to a certain point. Supplemental energy helps to use more of the protein contained in high-quality pasture forage. During early lactation cows need extra energy to return to positive balance for milk production and breeding back. During late lactation supplemental energy can help cows build body condition to a score of 3.5. (It's recommended that cows freshen with a body score of 3.0 to 3.5.)

Dry Matter Intake

Dry matter intake is extremely important in feeding dairy cows. Cows can eat more than 4% of their bodyweight equivalent in dry matter per day. How much of that daily dry matter (and its content of energy, protein, fiber, and other nutrients) intake need is met with high-quality pasture forage greatly affects the profitability of producing milk.

Research in Vermont in 1993 showed that it's possible to reduce grain supplementation so that cows increase pasture forage dry matter consumption to 3% of bodyweight per day on pasture (was mainly orchardgrass, Kentucky bluegrass, & 10-20% white clover), without loss in milk yield. The traditional pasture forage dry matter intake recommendation was 2% of bodyweight.

The questions of what the upper limit of pasture forage dry matter intake and actual protein and energy requirements are for cows producing more than 35 to 50 lb milk/day still need to be answered. This will require research that includes unsupplemented cows so that milk yield and economic responses to supplemental feeding can be determined for cows grazing well-developed, well-managed pasture. It also requires monitoring cows that weren't supplemented on pasture to determine their long-term (2-3 years) reproductive and milk yield performance.

If cows eat 3 to 4% of bodyweight in pasture forage dry matter per day, that amounts to 24 to 52 pounds of dry matter for cows weighing 800 to 1300 pounds. Since pasture forage contains about 25% dry matter, cows must eat 100 to 200 pounds of high-quality, palatable green pasture forage each day! Grazing management must make it as easy as possible for cows to eat that much.

We found that saving 15% of a paddock area for grazing 2 hours before milking increased pasture forage intake and milk yield. We gave cows 85% of a fresh paddock after the morning milking; then they got the remaining 15% 2 hours before the afternoon milking. This prompted intensive grazing of the high-quality forage, instead of the herd just standing for 2 hours at the gate waiting to go to be milked and fed supplements. This management change, which only involved daytime grazing, increased milk yield 2 lb/cow/day.

An extra break wire had to be put up and taken down in each paddock, so more labor was required. (An automatic gate opener could eliminate two trips to the paddock to open the extra break wire and the gate: e.g. Batt-Latch, 42 Bankwood Rd, Hamilton, New Zealand. FAX: 64-7-855-4482) We didn't decrease supplement level, but it's possible that it might be reduced without affecting milk yield of cows fed this way. The extra milk was worth about $9/cow/month. That could add up to real money!

Protein
Nitrogenous components of various kinds, including true protein and nonprotein nitrogen, are used in the protein metabolism of ruminants. True protein components can be divided into fractions that are either degraded in the

rumen, or undegraded and pass through the rumen (bypass protein). Most of the nonprotein nitrogen and rumen degradable fractions are broken down and used by rumen microorganisms in forming microbial protein. Generally microbial protein is used to meet protein requirements of ruminants, but it may not be adequate to satisfy the needs of dairy cows producing more than 35 (Jerseys) to 50 (Holsteins) pounds of milk per day. Protein that passes through the rumen undegraded can serve as a source of amino acids that can in effect supplement microbial protein to improve milk production.

Most of the protein and nonprotein nitrogen in high quality, lush pasture forage degrades in the rumen. Since the level of bypass protein needed for milking cows is thought to be about 37% of total protein, there may be special problems (or opportunities) in managing rations for dairy cows being fed on high-quality pasture. More research is needed, but as noted above, research in Ireland indicated that protein (including bypass protein) supplements may not profitably benefit cows grazing well-managed pasture. It all depends on forage protein content, cost of protein, and value of milk produced.

Many farmers unnecessarily feed protein supplements to cows grazing high-quality pasture, which can average 22% crude protein. If there's too much protein in the total ration, cows use energy to eliminate excess protein nitrogen; this energy could otherwise be used for production. Feeding straight corn meal to supplement energy, rather than a mixed pellet with even 12% crude protein, is best for cows grazing high-quality pasture, because corn meal doesn't add to their already excessive protein load (D. Hoke DVM).

Fiber
The more milk a cow produces, the more digestible fiber she needs in the total ration to sustain a maximum percentage of butterfat in the milk. But when the

proportion of fiber in the ration rises above a certain level, milk production drops. For each cow a point exists where milk production can only increase by decreasing the percentage of butterfat in the milk, and butterfat can be increased only by decreasing milk production. Feeding high-starch grains (e.g. corn, barley) may further depress milk butterfat content from cows grazing lush pasture.

Fiber in forage stimulates the rumen and increases formation of acetic acid, a precursor of milk fat. Fiber also promotes cud-chewing and secretion of saliva, which acts as a buffer in the rumen and favors production of acetic acid. Fiber may also slow passage of high-quality pasture forage material through the rumen, resulting in more complete digestion and use of its protein and energy.

To provide more fiber, some farmers have fed 1 to 2 lb of hay/cow/day to cows grazing lush pasture. Farmers who stopped feeding this supplemental fiber found that their cows grazed more and milk yield increased.

During many years of selecting cows (especially Holsteins) in the United States that are fed with large amounts of grain and concentrates, rumen development may not be what it should be, or cows may have been bred away from efficient use of pasture forage. Also, cows that have been fed corn silage during the several months of winter confinement may not be very well adapted to efficiently using pasture forage, no matter how well the pasture is managed. To achieve the highest production and profit from pasture forage, it might be better to feed cows haylage or hay, rather than corn silage during the off-season (e.g. late fall-winter-early spring).

Raising replacement heifers on well-managed pasture (beginning during their first week of life; described in this chapter), may eliminate rumen development problems that may be limiting efficiency of pasture forage use.

Body Condition
Cows normally change body condition as they store fat

and then use it to balance changes in energy demand. Peak demand of stored energy occurs shortly after calving.

Body condition scoring is a way of monitoring how well you're managing energy in your cows. Monitoring occurs over time, so body condition scores only have meaning when used to compare one moment to another in a cow's life. Having a good body condition score is not the same as being well fleshed. A cow has a good score when her body condition is optimal relative to timing of events in her life, cycle of events on the farm, and historic performance of similar animals on the same farm.

Incorrect understanding and use of body condition scoring has resulted in heavier cows, which has negative production and economic consequences. Large, fat cows compared to smaller, thinner cows:

- Lose more condition during the first 6 weeks of lactation. High fat reserves directly inhibit appetite, and metabolic effects of processing stored fat suppresses appetite, so fat cows have reduced dry matter intake.
- Don't live as long because of cumulative effects of metabolizing more stored fat.
- Have more reproductive (e.g. dystocia, infertility, metritis, retained placenta, vaginal prolapse, vaginitis) and digestive (e.g. abomasal displacement, fat cow syndrome, ketosis, pregnancy toxemia) disorders.
- Take more time to first ovulation and first heat.
- Have a lower first-service conception rate.
- Have more difficulty calving.
- Have higher incidence of milk fever.
- Fat heifers need more services per conception and produce below genetic potential.

Inadequate energy in the feed results in animals that are too thin, with these production and economic consequences:

- Thin cows have lower peak lactation production, higher dry matter consumption in first third of lactation, and

may not achieve genetic potential to produce milk.
- Thin heifers are slow to reach puberty and may be too small to calve at 23 to 25 months old.

To avoid health disorders, excessive changes in body condition shouldn't occur. These are indicators of how well energy in the feed meets reproduction and production needs of individual cows and herd:
- Body condition remains relatively constant during the last 3 months of gestation (indicates that cow was not in negative energy balance before calving).
- No individual cow loses more than a full score (about 100 lb) during the first month after calving.
- Average cow doesn't lose more than one-half score (about 50 lb) during the first month after calving.
- Cows regain half of what they lost during the first month by the time they finish the third month of lactation (Hoke).

Vermont Experience
To answer questions about body condition of pasture-based cows, Extension Agents scored 1200 cows for body condition on 25 Vermont farms, 2 to 4 times during the 1993 grazing season. The sample of cows and herds represented a broad range of hill- and flat-land farms, grazing experience, and pasture use. All dairies produced year-round, with one inclining toward late winter calving. The scoring scale was: 1 to 5 (emaciated to hog-fat). Target scores were:
- Cows:
  - Dry off: 3+ to 4-
  - Calve: 3+ to 4-
  - Mid- to late lactation: 3
- Heifers:
  - Growing: 3- to 3+
  - Calve: about 3+

Dave Hoke, D.V.M. studied the scores and concluded:
- 15 herds had no change in whole herd average score

from turnout to midsummer.
- 4 herds gained, and 6 herds lost in whole herd average score, but changes were not large enough in either direction to represent significant or perceptible change in the field.
- 22 herds had normal condition loss in early lactation during the grazing season.
- 2 herds showed a tendency for fresh cows to retain post-calving condition through early lactation.
- 1 herd showed a tendency to lose excessive body condition after calving.
- Mixed swards of cool-season grasses, legumes, and herbs that are in reasonable vegetative condition are seldom deficient enough in quality to produce dramatic changes in body condition, even in fresh and early-lactation cows, unless total dry matter offered is deficient, or plant density restricts dry matter intake.

Drying Off

Cows should be dried off at least 45 to 60 days before freshening. During the dry period, cows can be grazed on pasture containing less energy (30% over maintenance needs) than that shown in the tables in chapter 8. In fact, grazing a forage containing less energy may help to dry cows off and prevent them from becoming too fat. A good way to lower energy content of forage available to dry cows, is to have them graze paddocks after milking cows have eaten the highest quality forage.

Jersey dry cows may need to be removed from pasture a month before calving, to avoid milk fever. Even when dry cows follow milking cows to clean up paddocks, the forage may still contain too much calcium and too little fiber for Jersey dry cows.

Seasonal Dairying

Another way of drying cows off is to:
- Breed the entire herd to calve on pasture when plant

growth starts in the spring;
- Produce most of your milk as cheaply as possible during the grazing season that you extend as long as possible into the fall;
- Then dry the entire herd off: youngest animals first so they grow more, older cows next, then all others;
- Take a well-deserved vacation from milking (see economics of seasonal dairying in chapter 11).

Seasonal Breeding Of Dairy Cows

Seasonal dairying requires that you pay especially close attention to your cows' reproductive requirements. You need to keep complete, accurate breeding records to achieve and maintain a seasonal milk production pattern, and culling due to reproduction problems at a minimum.

The main concern is to have cows in proper body condition at calving, and lose minimal condition during early lactation. This is achieved by making certain that dry cow transition and early lactation nutrition is adequate. Proper body condition ensures that cows cycle and are fertile when the breeding period begins. If cows lose weight rapidly in early lactation, their best reproductive performance won't occur until after 120 days in milk, creating a problem for calving on a seasonal schedule.

The first step in planning is deciding which day milking will begin in the spring. Breeding should begin 280 days before that day. To allow first-calf heifers the extra time they need to begin cycling, their breeding should begin 10 to 20 days before cows. Of course, some heifers will calve and begin milking before the cows.

Sixty days after calving (beginning of heat detection period) all cows should have a reproductive exam by your veterinarian. Pay particular attention to cows that had retained placentas or uterine infections, so they can be treated if necessary before it's too late.

Begin careful heat detection 21 days before the breeding period starts. Record when cows came in heat and when they are expected in heat. Besides close observation, using tail paint greatly increases accuracy of heat detection. An interior waterproof, high-gloss, bright-colored enamel paint works well. One quart paints about 80 cows.

To achieve an absolute minimum 67% conception rate, you must:
• Detect 90 to 100% of cows in heat during the 21-day heat detection period before breeding starts;
• Have 90% of cows bred at least once during the first 21 days of breeding;
• Have 100% of cows bred at least once during the first 42 days of breeding;
• Get 60% of cows pregnant within the first 21 days of breeding, and 85% of cows pregnant within the first 42 days of breeding.

Breeding with bulls reduces the labor needed during the breeding season, but the problem of having bulls on the farm and possible milk yield loss in future generations from using bulls of unknown genetic merit must be considered. A way that works well is to use bulls only after one or two breeding cycles with artificial insemination. You need one bull per 25 cows for young bulls, or one bull per 35 cows for older bulls. To be certain all cows are bred, at least 2 bulls should be used, even for a herd of 25 cows.

If you raise your own replacement heifers, it may be advantageous to use semen of different price and genetic merit for different times of the breeding period. During the first 21 days use the most expensive semen, during the next 21 days use moderately priced semen, and use the cheapest semen after that. The reason is that replacements need to be reared from calves that are born earliest, and this method ensures that they will be sired by the best bulls. If you sell all of your calves, it may be advantageous to discuss which quality of semen to use with your calf buyers. The quality of semen used should be reflected in the price you obtain for your calves.

You can concentrate breeding and calving within

shorter periods without affecting fertility, by synchronizing estrus with prostaglandins (e.g. Lutalyse, Estrumate). A practical way to synchronize breeding is to:
• Breed all cows that come into heat during the first 6 days of the breeding period;
• On the 7th day inject with prostaglandin all cows that weren't bred during the first 6 days;
• Observe for heat and breed cows in heat during the next 6 days;
• On the 7th day again inject with prostaglandin all cows that aren't yet bred, and breed when you detect heat; and
• Have any cows not yet bred examined by your veterinarian.

About 35 days after insemination your veterinarian should palpate cows to determine if any are open. These can be rebred and sold to other dairy farmers, or induced to calve early. Be aware that inducing early calving may result in small, weak, or dead calves, decreased milk yield in that lactation, and retained placentas (Brickner).

Milk Fever

Dry dairy cows are a special group that should be managed separately during the last month of their dry period, to avoid having them come down with milk fever when they freshen. Recent research indicates that milk fever results from high potassium content of feed, which lowers blood calcium level of cows eating it. A cow has large stores of calcium in her skeleton and plenty in the feed in her digestive tract. But if she only has a small amount circulating in her blood, it isn't enough to meet the drastic change from the needs of the developing calf, to the demand of milk production in early lactation. The deficiency is only temporary, however, because the cow has ample reserves of calcium. A cow suffering from milk fever can't mobilize calcium reserves quickly enough. A few days after calving the rate of calcium mobilization from reserves increases, and she is then able to cope with the demands of lactation.

Calcium flows from the blood into the milk, and unless blood levels are replenished from reserves in bones and intestines quickly enough, blood calcium level falls, and milk fever results. If untreated, muscle paralysis occurs and the cow rolls over on her side, and is unable to get up again. When she is on her side, rumen gases can't escape, so she becomes bloated and dies from either pressure on her heart or from inhaling rumen contents. Treatment requires a timely injection of calcium and care by a veterinarian.

Milk fever mainly occurs in second-calf heifers and older cows. Older cows are much more susceptible because the calcium reserves in their skeletons are less available. Jerseys are more susceptible than other breeds, because of the high demands of lactation in relation to their size and total metabolism. Milk fever usually isn't a problem with first-calf heifers.

Spring- or autumn-freshening cows grazing lush pasture may develop milk fever, unless managed appropriately. Lush well-managed pasture is not good feed for dry cows during the last month (especially last 3 weeks) of their dry periods, because potassium content of the forage is too high and fiber content is too low for their needs. During the last month of their dry periods, cows should be fed a low potassium, high fiber ration; their calcium mobilization process will then be active enough at calving to cope with the increased demand.

During the last month of the dry period, rations should have the following characteristics: 12% crude protein, 35% acid detergent fiber, 0.5% calcium, 0.4% phosphorus, and 0.60 Mcal/lb net energy lactation. They should be fed long-stem dry hay to keep their rumens full and stimulate good rumen muscle tone. A few days after calving, fresh cows can go back on pasture. (G. Catlin, VMD and J. Kunkle, VMD)

Tail Docking

This practice was imported from New Zealand; it just goes to show that not everything from there is a good idea. I've been swished in the face by cows many times, but it never occurred to me to cut their tails off!

Removing the tails of dairy cows is a bad idea. If it only was used as a fly swisher it still would be worth it to the cow to keep it. Imagine never being able to get rid of flies crawling on and biting your back! Besides swishing flies, a cow uses her sensory tail to monitor the blind spot behind her, and it also helps balance animals when running.

If a tail is injured severely enough to require amputation, it should be done with anesthesia and sterile surgical technique. Blood supplied to the tail returns by way of a low-pressure venous sinus. Indiscriminate docking by strangulation or crushing in adult animals may result in chronic infection that continually seeds the circulatory system with abscess-forming bacteria, resulting in ongoing health problems, more veterinary expenses, and reduced production (Hoke).

We recently studied cow grazing behavior during several days on pasture, observing and recording what cows were doing every 15 minutes. As a result, we have photographs of cows with docked tails going through all kinds of tormented contortions to get flies off of them. This included throwing mouthfuls of grass on their backs! At the very least, that grass would be better for the farmer if it went down the cows' throats, and the time and energy spent struggling to reach the flies would be better spent in grazing and ruminating. These cows were treated for fly control, but the flies apparently weren't aware of it.

Tail docking is <u>very</u> bad from a public relations standpoint. I've noticed that urbanites hate to see cows with docked tails. These usually include people concerned about animal welfare. They also are the consumers of dairy products; they must not be alienated. It's only a matter of time before the urban public realizes what dairy farmers are doing to cows, and action is taken (e.g. milk boycott) to end the practice.

Rearing Calves on Pasture

Rearing calves tied up or loose in filth, covered with flies, in dark, dreary corners of poorly ventilated barns, or tied to individual hutches outside like mistreated dogs, does not produce the kind of adult cow needed for highest production and profits on pasture (or in confinement).

Both of these widely used calf-rearing practices only make work for farmers and provide income to veterinarians and calf hutch and feed dealers. They also result in unhealthy adult cows that don't know how to graze well or behave normally in the social organization of a herd, and may not be able to efficiently use the forage that you work so hard to produce. The amazing thing is that the calves being reared under such poor conditions are the dairy farmer's future!

There is a better way. Replacement heifers can be reared in groups on pasture (or in clean, well-ventilated, sunlit pens for confinement operations)! The only way to get healthy cows that have fully developed rumens and know how to graze well in a herd is to rear them from birth in groups on well-managed pasture.

Heifers raised on pasture tend to have larger rumens, which enable them to hold and process more feed and produce more milk than those raised in confinement. The relative size of a calf's rumen at 3 to 4 months of age remains constant for the rest of its life.

Calves can live on pasture forage alone after they are about 60 days old -- if they have been started on well-managed pasture during their first week of life. This is much easier and cheaper, and results in healthier, more efficient animals than raising them individually in barns or tied to hutches.

According to Paul McCarville (see References), modern farming pressures tend to cause the average farmer to disregard the physiology of the calf's digestive system. Understanding this system and providing for its requirements are essential for rearing calves successfully to be healthier and more efficient users of forage:

- The fourth stomach (abomasum) is the main digestive organ in a ruminant's early life. At birth, the abomasum is 3 to 4 times the size of the first stomach (rumen).
- When calves (or other ruminants) suck milk from their mothers, their heads and neck are in a low position that causes a groove at the bottom of the esophagus to form a tube that shunts milk directly to the abomasum, bypassing the first, second, and third stomachs. When calves are fed from buckets with their heads up and don't have to suck hard, the tube doesn't form fully and some milk gets into the rumen where it is wasted, and may interfere with rumen development. For this reason, calves reared separated from their mothers must be fed from nipples placed 18 inches (first 1-2 weeks) to 24 inches (2 weeks to weaning) from the floor or ground level.
- When sucking, a calf eats about every 2 hours, and each time it eats a clot of milk forms in the abomasum. By the end of the day, the abomasum is packed full of small clots that are acted on by digestive juices.

Setting Up And Doing It:
- Order nipples from:

    Paul McCarville, 18875 McCarville Lane, Mineral Point, WI 53565; Phone: 608-987-2416.

    Nate Leonards Pastures Unlimited, PO Box 490, Little York, NY 13087-0494; Phone: 607-749-3931.

    Nipples come with clear instructions. Complete sets of nipples installed on tubs, barrels, and trailers with barrels are available.
- Collect colostrum and, if necessary, freeze it until needed to feed newborne-to-week-old calves; never feed calves milk that contains antibiotics.

- Feed milk, rather than milk "replacer". Milk is a live food that contains essential enzymes, cellular materials, and nutrients that promote good health. If you're raising calves without cows, it's better to buy colostrum and milk from a dairy farmer than to use "replacers".
- Always have fresh, clean, cool water available. Calves need water in the morning more than they need milk.
- Calve on clean pasture, if possible, to prevent disease.
- Newborn. Remove calves from mothers, disinfect navels, and bottle feed 2 to 4 quarts of colostrum per calf within 6 hours of birth (and during each of the calf's next 2 days of life).
- 3 days old. Combine with other calves in groups of 4 to 7 in pens (inside or outside with shelter). Groups should consist of calves that have the same drinking speed, not necessarily the same age. Feed 2 to 4 quarts of colostrum per calf per day with a 5-gallon plastic bucket that has 5 to 9 nipples spaced around it near the bottom. Hang or place the bucket so the nipples are 18 inches from the ground. You'll have to teach calves to suck the nipples; allow a calf to suck your finger while you slip a nipple into its mouth. Leave buckets or barrels and nipples with the calves so they can suck them whenever they want, so they won't suck each other.

  Begin feeding calf starter supplement free choice at a rate of 1 to 2 lb/calf/day. Provide this in the evening so birds don't get eat it or contaminate it with manure from their feet.
- Calves always need to be sorted by drinking speed and placed with others that drink at the same rate.
- 5 days old. Combine in groups of 12 in outdoor pens on pasture (if weather permits, always with shelter). Feed each group from a cut-off half of a 55-gallon plastic barrel that has 15 nipples spaced 4 inches apart near bottom.

  During this stage, nipples must be placed close to bottoms of buckets and barrels, so that the milk flows easily to the very young calves. Both buckets and half-barrels must be placed so that nipples are 18 inches from the ground, allowing calves to suck with their heads in the correct position.

- Begin to train calves to respect electric fence. Place calves in a pen made with wooden sides or wire mesh fencing on pasture. Place three energized wires around the inside of the pen at 10, 16, and 28 inches from the ground on insulators or fiberglass posts. Some farmers successfully train calves with only one strand of polywire or polytape at calf nose height around the inside of the pen. Energized flexible net sheep fencing also can be used to make pens and train calves to respect electric fencing.
- 7 days old. Gradually begin increasing the amount of milk fed to 6 to 8 quarts/calf/day.
- When calves are sucking well, combine them with others in small paddocks to form groups of 12 that now drink from a barrel with a nipple-hose combination that results in slower milk consumption. Calves also have to suck harder to get milk, producing large amounts of saliva, resulting in healthier calves that don't scour.

Each of these groups of 12 calves now drink from 55-gallon plastic barrels that have 15 nipples spaced 4 inches apart around the barrels, 24 inches from barrel bottoms. Nipples have 24 inches of 7/16 outside-diameter plastic tubing inserted in them inside to reach barrel bottoms.

Barrels can be either stood on end in pen or pasture, or used placed on their sides on wheels. If used sideways, nipples need to be placed lengthwise around 35-gallon barrels, so that the nipples will be 24 inches from the ground, taking into account the height of the trailer or whatever else is used. Barrels on wheels can have 19 nipples spaced 6 inches apart to feed 15 calves. With barrels used sideways, plastic tubing needs to be only 18 inches long.

If you use barrels standing on end, it's best to haul milk to the calves in another container and pour it into the nippled barrels in the pasture. If you use a barrel on wheels, the whole contraption can be pulled to the milking parlor to be filled, and pulled back to feed the calves on pasture.

- Clean buckets, barrels, and nipples only after you're done using them for the season; the smell of milk and mother cows is needed to keep calves bonded to the equipment. If you use milk "replacer" you'll have to clean the equipment more frequently than once a season.

- By the time calves are placed in groups that suck from nipple-hose combinations, they should be well trained to respect electric fencing, and can be placed in small paddocks of 50 to 75 feet on a side. Paddocks can be formed with flexible net sheep fencing or 1 to 3 strands of polywire, depending on how much they respect fence.

- 4 weeks old. Feed as much warm water as calves want to drink in the morning; decrease milk allowance to 3 to 4 quarts of milk per calf per day and feed it in the evening.

- 6-7 weeks old. Decrease milk allowance one-half or, if calves are eating 2 lb calf starter/calf/day, they can be weaned from milk at this time.

- 6-9 weeks old. Wean by weight, grazing activity, and available pasture. Weaning time for calves on pasture varies, depending on how the calves are doing. Some farmers gradually decrease the amount of milk fed; others abruptly stop feeding milk.

- Treat for parasites as described under Parasite Control.

Grazing Management

The sooner you get the calves out of the barn and on to well-managed pasture, the better for the calves and you. Except for the first 3 days of bottle feeding, all training of calves to nipples and nipple-hose combinations can be done on pasture.

During the first 2 weeks that calves are on pasture, move them to a fresh paddock every time the forage is grazed down to about 1400 lb DM/acre (2 inches tall), or at least once a week. After 2 weeks, give the calves a fresh paddock every 1 to 2 days. Pregrazing mass should be about 2100 lb DM/acre (4 inches tall).

Calves should be allowed to leave a high postgrazing mass (1400 lb DM/acre) and not forced to graze close to the ground. Calves graze high if given adequate forage allowance. This helps prevent them from ingesting parasite larvae, which tend to be in the bottom 1 to 1.5 inches of the sward.

Calf pasture is best kept in good condition by grazing it down periodically with cows, rather than by mowing, which leaves stubble that irritates calves' mouths. Periodic grazing by cows also helps transfer needed rumen microorganisms to calves.

Take advantage of feeding times to move calves into fresh paddocks and set up fences. They'll go wherever the barrels of milk go, and while they drink milk you can move fences without having calves getting out.

Maintain calf paddock fencing energized at all times except when you're moving it while they're drinking. If they figure out when the fence is off, they will get out, and you'll learn that dairy calves can outrun humans.

One farmer I know, turned all of his fence off when he would bring the cows in to milk or move cow paddock fencing. His calves figured out what he was doing, and soon every time they saw him going toward the cows, the calves broke out of their paddock. It's very difficult to retrain calves to respect electric fence once they've tasted freedom. He had to tie calves to tractor tires and to each other, to slow them down until they were retrained. He now leaves calf paddock fencing energized permanently. He's also in better physical condition now, from running after calves. Don't chase calves if you aren't in good condition; it's not worth having a heart attack over loose calves! It's much easier to just entice them back into the paddock at feeding time.

Supplements

New Zealanders feed 0 to 2 lb grain supplement/calf/day to calves being fed milk or milk replacer on well-managed pasture; at 12 weeks old, their calves live on pasture alone. I bet that we can do the same on well-developed pasture.

American farmers generally feel that their pasture forage quality isn't adequate for good calf growth, so they supplement their calves. In an experiment with weaned Holstein and crossbred calves given a fresh break of very good Kentucky bluegrass/white clover pasture every 3 days, we found that feeding 2 lb supplement/calf/day (Blue Seal calf starter: 19% crude protein with Bovatec for control of coccidiosis) resulted in the same weight gain (1.6 lb/calf/day) as feeding 4 lb. This summer (1997) we did another experiment with crossbred calves fed 0 or 2 lb/calf/day of the same supplement after weaning. Calves were given a fresh break of fair Kentucky bluegrass/white clover pasture every day. Supplemented calves gained an average of 1 lb/calf/day; unsupplemented calves gained 0.6 lb/calf/day.

Be careful in feeding supplements to calves or any other livestock. Make certain, as best you can, that you're not simply substituting expensive supplement for inexpensive pasture forage. Be especially careful with heifer calves so that they don't gain weight too fast. Irish research showed that high rates (greater than 1.5 lb/day) of heifer weight gain before sexual maturity (about 8 months old) slows mammary gland development and reduces later milk production.

Always have available salt and any minerals needed in your area. It may be best to use feeders that meter minerals out through the water system, to be certain that calves get enough minerals.

Growing Calves and Heifers

Weaned calves can either be set-stocked at 2 to 3 per paddock and lactating cows rotated through the paddocks, or grazed as a leader group (mob) ahead of the cows. Set-stocked young calves are territorial and usually don't follow cows out of home paddocks. Calves in leader mobs get better feed and tend to get less parasites than if they follow cows and have to graze close to the ground where parasite larvae concentrate.

It's best to monitor (weigh or tape) heifer weights so you can manage them as needed. If leader-grazing heifers gain more than 1.5 lb/day before they're 8 months old, they should be changed to follow cows. Also be careful to control the residue left by cows so that following heifers don't gain more than 1.5 lb/day.

Heifers should be fed to breed at 13 to 15 months old, so they calve when they are 22 to 24 months old. After heifers are bred they can be treated the same as dry cows, cleaning up after lactating cows, so the heifers don't get too fat. Thirty to 45 days before calving, heifers should begin grazing the same high-energy forage as milking cows receive.

Growing male dairy stocker calves need high quality forage to achieve the high rates of weight gain (2.2 lb/calf/day or 1000 lb/acre/6 months) that are possible. Grain supplements probably won't be needed if you manage grazing so pregrazing mass is 2100 (4 inches tall) and postgrazing mass is 1400 (2 inches tall).

Parasite Control
Calves and yearlings become infected with internal parasites when they graze contaminated pasture. Since young stock have little resistance, the parasites multiply greatly in the animals, and can slow their weight gains and growth. In leader-follower grazing, heifers grazing in front of or behind cows actually can contaminate the cows with parasites in the next rotation, because of large amounts of parasite eggs shed by the heifers. Although cows develop resistance, heavy parasite loads adversely affect cows' appetite, milk yield, and reproduction.

Controlling parasites in cows during the first 100 days after calving has resulted in 100 to 1000 lb more milk produced per cow. Parasite control in heifers has increased weight gains by up to 107 pounds during the grazing season, and decreased time to calving by more than 2 months.

The period between grazings isn't long enough to kill parasites, so other management practices are needed to reduce contamination of pasture and infection of all livestock:

- Alternate pasture and hay or haylage crops in spring and early summer; i.e. an area is grazed one year, and machine-harvested the next year.
- Graze calves without adults only on clean pasture (e.g. grazed by other species or machine-harvested during previous year, not grazed by adults of the same species before in current year: see page 114, or new seeding). This can be as effective as treating the animals with a dewormer.
- Graze mixed species of animals; e.g. sheep with cows. Parasites of one specie usually don't affect others, so each specie can clean up parasites of others without harm. This is another example of why diversity in plants and animals is beneficial.
- Harrow pasture to break up dung pats during hot weather. This kills parasites by drying them. (Note: parasite contamination increases if pasture is harrowed during cool, moist weather.)
- Deworm at strategic times. This requires multiple treatments with a parasiticide such as fenbendazole (e.g. Hoechst-Roussel Panacur or Safe-Guard) at certain intervals during the grazing season. Multiple, spaced treatments are needed because of the time it takes parasites to complete life cycles in young stock, depending on their size and age. Parasites take longer to reach maturity in adult cows.

Adult cows: deworm in late fall after first hard frost, and again 6 weeks after start of grazing season in spring.

Calves & Heifers:

Less than 300 to 400 lb weight: deworm 3 to 4 weeks

after turnout on pasture and again 3 to 4 weeks later.

400 to 800 lb weight: deworm at turnout, 3 to 4 weeks after turnout, and 3 to 4 weeks later; treatment at turnout isn't needed if animals were dewormed at the end of the previous grazing season.

Greater than 800 lb weight: deworm at turnout and 4 to 5 weeks later; treatment at turnout isn't needed if animals were dewormed in previous fall.

First-calf heifers: deworm before they enter adult herd.

• Monitor parasite levels by fecal sampling and analysis. For best results collect at least two samples from each phase of herd: 3- to 6-month-old calves, open heifers, bred heifers, dry cows, and fresh cows. Collect walnut-sized samples from fresh dung. An easy way is to use inverted zip-lock bags over your hand to pick up samples from the tops of dung pats. Squeeze air pockets out of bags before sealing to prevent eggs from developing. Mail in a styrofoam cooler with cold packs and a check ($3/sample) to: Dr. Gil Myers, 3289 Mt. Sherman Road, Magnolia, KY 42757. The samples will be examined and a report mailed to you within 1 week of receipt.

Summary Advice For Dairy Farm Maximum Profit
Visitors from Ireland and New Zealand who observe American dairy production methods, consistently have offered well-intentioned, helpful advice.

For example, Irish dairy pasture researcher Sinclair Mayne has suggested these ways for American dairy farmers to increase profit:

• Grain supplement should never be more than 12 to 15 lb/cow/day; feeding more just substitutes for pasture. It's best to feed grain only once a day, after evening milking.

• Always feed grain according to milk yield, never at a fixed rate to all cows.

• Feeding silage (corn silage or haylage) makes cows lazy grazers and reduces pasture forage intake. Silage should

only be fed when pasture is in short supply and extra feed is needed to slow the rotation.
- Have a short calving period, beginning 1 month before start of grazing.
- Turn cows out as early in spring as possible, even if for only 2 hours per day at first.
- Decrease silage and grain feeding as quickly as pasture growth increases.
- Monitor pasture growth by walking all paddocks once a week and budget feed supply accordingly.
- Conserve spring pasture surplus as haylage to keep plants in vegetative stage of growth.
- World price competitive production requires:
  - maximum use of grazing
  - strategic use of grain
  - aggressive foraging cows
  - minimum reliance on housing
  - maximum reliance on contracted machinery work

  New Zealand feed planner Ian Brookes boils it all down to:
- Grow as much high-quality pasture forage as possible.
- Use as much of the pasture forage grown as possible.
- Feed each animal as much pasture forage as possible.

BEEF CATTLE

One of the most disgusting things that I have ever seen and smelled was a beef cattle feedlot off of Interstate 80 in Nebraska. I'm sure that if people knew their beef was coming out of cesspools such as that, less beef would be eaten. Feedlots now produce more organic waste than the total amount of sewage from all municipalities in the United States! Feedlots operate at a profit only because they are allowed to pollute the environment and pass those costs on to society, which sooner or later has to pay them in some way. It's another example of economics as if people, animals, quality of life, and the Earth don't matter.

Feedlots really can't even be justified on economic

terms. Feed certainly affects total productivity and profit in beef operations, accounting for about 75 percent of the cost of keeping cows, and raising and finishing stockers. But grazing them on pasture can cost only about one-sixth as much to accomplish those objectives than it does in confinement! And that doesn't include the savings of about 2 bushels of topsoil that erode away in producing each bushel of corn, which is used to "finish" animals to the point that their meat is too fat and causes health problems in people that eat it. I wonder how much money is spent each year in this country just trimming fat from feedlot-finished carcasses.

It now turns out that pastured beef is good for people! Irish researchers recently showed that beef from cattle finished on pasture had a proportion of omega-3 fatty acid four times greater than in beef from animals finished in feedlot. A high level of omega-3 fatty acid in people's diets has been found to be beneficial for health.

Pasture Finishing
If you think that feedlot finished beef tastes good, you haven't lived! The best beef I've ever eaten was in a Buenos Aires restaurant. I didn't need a knife to cut the meat; I used a fork! In Argentina pasture finishing is the most common way to fatten animals.

Almost all livestock in Argentina are raised and fed on pasture forage alone. Pastures consist of the same grasses and legumes used in the USA (e.g. orchardgrass, bromegrass, phalaris, tall fescue, white clover, alfalfa). Livestock graze rye, oats, or triticale forage during winter, and corn or forage sorghum in midsummer.

The same kind of finishing can be done here at a cost of 20 to 30 cents per pound of gain (feedlot finishing costs 45 to 50 cents per pound of gain):
• Need high quality pasture such as Kentucky bluegrass, perennial ryegrass, orchardgrass, bromegrass, or endophyte-free tall fescue, with 30 to 50 % legume.

- Pregrazing sward 6 to 8 inches tall; finishing cattle graze first down to 3 to 4 inches tall; brood cows follow, grazing down to 2 inches.
- Finishing cattle need to be moved to fresh pasture every 1 to 2 days to achieve required gains of 2 lb/day.
- As long as plant growth is adequate, don't substitute grain for pasture forage.
- If you feed grain, vary the amount with the season:
  - Spring, when pasture quality is high, limit grain feeding to 0.5 % of bodyweight.
  - Midsummer, when forage quality declines because of rising lignin content, increase grain feeding to 1 % of bodyweight.
- If you feed grain at more than 1 % of bodyweight, divide it into more than one feeding to avoid acidosis.
- Shelled corn works well; no protein supplement is needed if pasture forage contains more than 16 % crude protein.
- Feed grain on the ground under an electric wire; loss is less than 5 %.
- Feed grain after the morning graze, between 12 and 4 p.m.; this increases rate of gain and feed conversion up to 20 %.
- Pasture finishing works best with cattle that finish at 1050 to 1200 lb.
- Yellow fat in animals finished on pasture is due to legumes in the forage. So to get white fat, graze animals during the last 4 to 6 weeks on nitrogen-fertilized grass.
- Market cattle by 24 months old to ensure tenderness (Bartholomew & Martz).

Cow/Calf And Stockers

The faster beef animals grow the more efficient the production is, in terms of forage dry matter, protein, and energy needed, because the daily effect of the maintenance requirement on total production is reduced. For example, 1,200 days, 16,140 pounds of dry matter, and 1,438 pounds of crude protein are needed for a steer gaining 0.55 pounds per day to grow from a 350-pound starting weight to a 1,000-pound market weight. In contrast, a steer gaining 1.65 pounds per day reaches the 1,000-pound market weight from a 350-pound starting weight in 400 days, and only needs 6,730 pounds of dry matter and 684 pounds of crude protein to do it!

High rates of daily weight gain are possible to achieve in other ways than confining animals in feedlots. Feed-cost savings can be very large when beef cattle are grazed on intensively managed pasture, because daily weight gains of 2.75 to 3.3 pounds are possible.

Feeding beef cattle as much high-quality forage as they can eat also results in large savings of pasture forage if rapid growth is maintained. This allows more animals to be carried on a pasture, and increases the profitability of the operation.

Supplements

Irish researchers found that energy and protein supplements are unnecessary when beef animals graze green pasture. As long as sward height (i.e. 6 inches pregrazing, 3 inches postgrazing after producing animals) and quality are adequate, unsupplemented pasture forage is the most profitable way to produce beef.

The research showed that:
- Supplementing stocker cattle grazing well-managed pasture didn't affect rate of gain.
- With stockers it was cheaper to use subsequent compensatory gain rather than feed supplements to maintain high rates of gain during pasture shortages.
- Supplementing stockers during pasture shortages reduced subsequent compensatory gain (Mayne).

Weaning Calves

It's best to wean calves on the same high quality pasture forage that they've been grazing with their mothers.

Calves are stressed less if they experience the least amount of change at weaning. So the only thing that should change is that calves can't nurse their mothers.

One way to accomplish this is to put up a 3-wire electric fence between where the herd is grazing and the next paddock break. Separate mothers from calves, either in a corral or on pasture by moving cows with a polytape between two people. Place cows in the break where they were and place calves in the new break, separated from the cows by the 3-wire fence. The calves graze the fresh pasture and don't attempt to return to their mothers. Continue moving the two groups, separated by 3-wire break fences, through the pasture in the normal sequence for 2 days. Then you can move the calves further away, preferably out of sight of the cows (Salatin). After a few days, cows can follow near calves to graze down paddocks.

Move weaned calves every 1 to 3 days to fresh paddocks containing pregrazing pasture masses of about 2100 lb DM/acre (4-6 inches tall). Calves should graze down to about 1400 lb DM/acre (3 inches tall), followed by dry cows to graze down to 1000 to 1200 lb DM/acre (1-2 inches tall).

Parasite Control
Internal parasites can reduce weight gains by up to 70% in calves and yearlings, so it certainly pays to manage in ways that decrease pasture parasite contamination. The same practices described above to reduce parasites in dairy calves and yearlings can be used in beef production.

Partnering
If you partner (joint venture) the raising of calves and stockers with someone in another region of the country, who has pasture forage at an opposite time of the year from your area, you both can win. For example, a farmer in Mississippi can raise calves to about 550 pounds from birth during October to April on annual ryegrass pasture.

The calves are shipped to you in Vermont to begin grazing your pasture on May 1. You raise them to 800 pounds during May to October on intensively managed permanent Kentucky bluegrass-orchardgrass-white clover pasture. (See chapter 8 and Allan Nation's *Pasture Profits with Stocker Cattle* in References.)

Cows

Beef cow (and other female livestock) nutritive requirements depend on their size, age, and stage of reproductive cycle. Their requirements change markedly, especially at birthing and weaning. Generally cows need an increasing level of nutrition before birthing, a high level during lactation, and lower nutrition after weaning. If you're flushing the cows, a high level of nutrition is needed at least 30 days before the beginning of breeding.

Be careful in the feeding of beef cows so that they don't become too fat grazing intensively managed pasture or eating the high quality surplus hay or haylage that is harvested. If they're too fat at breeding time, they may not conceive.

One way to have cows gaining weight at breeding time is to begin grazing them behind their calves, as soon as the calves are weaned in late summer or autumn. This will lower the nutritional level of the cow's pasture forage intake and help keep them from becoming too fat. It will also increase the nutritional level of the forage that the calves are able to eat, since they won't be competing with the cows. Overwinter the cows on hay that provides a maintenance level of nutrition.

In addition to the information shown in the tables in chapter 8, it should be noted that dry pregnant mature beef cows need about the same amounts of energy as those shown for dry dairy cows. Beef cows nursing calves need 3 to 5 Mcal more energy per day than they needed before calving.

Replacement Heifers

Growing replacement heifers can be grazed on forage containing somewhat less energy than that needed by stockers or cows nursing calves. Feeding too much energy to heifers before breeding results in them producing less milk as cows, so be careful. One way to keep heifers on a lower plane of nutrition is to have heifers follow leader grazers (e.g. cows with calves or stockers) through paddocks to eat remaining forage that contains less energy.

First-calf heifers also need to be managed carefully to avoid conception problems. Research has shown that 81% of first-calf heifers breed back and calve during the first 30 days of calving season, when their first calves were weaned at 56 days. If their first calves were weaned at the traditional 240 days, only 46% of the heifers bred back.

SHEEP

Mature sheep spend almost their entire lives at maintenance level, which means that their energy needs are more or less constant, except during certain periods in gestation and lactation. Liveweights of ewes fluctuate during periods when feed quality is adjusted to the animals' needs. The energy requirement of adult ewes during the last 4 weeks of gestation is 1.5 times their maintenance need. This increases to 3 times maintenance during early lactation, and decreases to 2 times maintenance by the third month of lactation. After weaning, ewes can return to maintenance, until flushing.

During maintenance periods ewes can tolerate forage with lower energy contents, so they can be used during such times to clean up paddocks after lambs or other livestock, or in mob-stocking of rough land for pasture improvement (see chapter 10). Be careful in using mob-stocking so that ewes aren't kept on poor-quality feed for too long, though, or a reduced number of lambs may be born the next spring. It's best to alternate a day of grazing low-quality pasture with a day of grazing higher-quality pasture, to avoid stressing the animals too much. Grazing ewes on forage that is better than their minimal need results in them weighing more and consistently giving birth to more and larger lambs that gain weight faster.

Lambs

For a certain rate of liveweight gain, lambs need a greater energy intake if they are grazing low-quality pasture, compared to that needed when grazing high-quality pasture. Fast rates of gain can't be achieved with low-quality pasture, because the bulk of feed in the rumen limits intake before enough energy has been ingested to reach high weight gains. Therefore, the energy of low-quality pasture is used less efficiently than that of high-quality pasture. This difference between low- and high-quality pasture becomes more obvious as daily weight gain increases, in terms of energy requirements for good performance.

Besides saving a great deal of work for shepherds (No more sleepless nights attending lambing jugs. Yes Virginia, sheep can give birth unassisted by humans!), lambing on pasture in spring enables ewes to be fed on high-quality forage during most of their lactation, resulting in considerable savings from not having to feed concentrates to produce the milk. Lambs begin to graze soon after birth on pasture, and the forage they eat stimulates rumen development, so they progressively make more use of forage.

Successful spring lambing on pasture requires ewes with good mothering instincts. Shed lambing with intense shepherd care covers many weaknesses in sheep that don't become apparent until they lamb on pasture with less care. Good mothering breeds include Romney Coopworth, Corriedale, Rambouillet, Targhee, Columbia, Border and North Country Cheviots, Perendale, Montadale, and their crosses.

For a complete description of New Zealand style

lambing on pasture, do yourself a favor and buy Don and Virginia Wilkinson's Pasture Lambing video for $39.95 from Red Hill Associates, 3038 Red Hill Road, Oakland, OR 97462; Phone: 503-849-2907. The video provides a detailed overview of the advantages, equipment, and steps needed to prepare a farm for successful lambing on pasture.

Lambs have the fastest rate of gain on pasture while nursing their mothers' milk. Pregrazing pasture mass should be about 2100 lb DM/acre (about 4 inches tall) for ewes with lambs or lambs alone. Postgrazing pasture mass shouldn't get below 1200 lb DM/acre (1.5-2 inches tall) for best milk production and lamb weight gain.

Ideally, lambs reach market weight at 14 to 16 weeks of age from mother's milk and pasture forage, without any grain or concentrate feeding, and can be taken directly to market at that time. Lambs that haven't reached market weight or are being kept as replacements can be weaned and separated from their mothers at 14 to 16 weeks of age.

Parasite Management
Lambs may need to be dewormed at 4 and 8 weeks after turnout, to kill parasites that they picked up while grazing with their mothers. Ewes should be dewormed right after lambing. Analyze fecal samples to see if treatment is needed.

It's best if weaned and dewormed lambs graze clean pasture that was set aside in the spring and machine-harvested or grazed by other livestock species. Lambs then enter a pasture area that wasn't grazed by sheep that would have shed parasite eggs during the current season; the lambs should have less parasites for most of the season.

If lambs can't be placed on relatively clean pasture after deworming at weaning, they may need to be dewormed twice more at 21-day intervals, then every 28 days until the end of the grazing season or until they're sold.

In Vermont, we found a way to manage parasites without chemicals that may work in other northern areas to keep parasite levels low in lambs (see page 114). Start with pasture not grazed by sheep in the current year. Graze each paddock with lambs and ewes only once per season.

Other possible ways for controlling parasites in ewes

and lambs on pasture besides individually treating animals with anthelmintics are:

- Alternate areas grazed by sheep with those grazed by cattle, either within a season or preferably in different years. Swapping sheep and cattle pastures on July 1 lowers parasite levels in both animal species.
- Companion-graze beef cattle and sheep. Sheep can be grazed ahead of, with, or behind the cattle. Besides reducing parasite levels in both species, this results in decreased pasture weeds, better total pasture use, and improved sheep performance.
- Feed diatamaceaous earth mixed half and half with salt.

No matter which method you use to control parasites, it's best to monitor your flock by checking dung samples for worm eggs, to be certain that parasite levels are low.

--∞--

All of the above practices ideally promote high, sustained daily weight gains of lambs, while minimizing production costs and maximizing profits. Unfortunately, wholesale lamb markets in the United States require a finish and size of lamb that may be difficult to obtain on pasture forage alone, given the sheep that commonly exist here. Achieving 16-week-old lamb liveweights of 85 to 90 pounds on pasture without grain or concentrate supplement is relatively easy to do. But it's more difficult to consistently get lambs up to the 105- or 110-pound liveweight required by current markets, without supplements.

Whole corn or some other energy supplement can be fed to lambs on pasture, beginning just before the end of the grazing season, but it doesn't make any sense. Judging from French market requirements, the ideal lamb slaughter liveweight range is 80 to 88 pounds. This is because at that weight the meat is more lean, tender, and tasty. As lambs are raised to 105 pounds on grain, they do get bigger, but they also get fatter and less tender. Who wants to eat fat lamb? This requirement for bigness may disappear as producers and consumers realize the benefits of lean lamb grown entirely on pasture without grain.

Obviously, this realization will have to be preceded by more widespread development and use of suitable high-quality pasture, along with appropriate selection and breeding of sheep. The large long-legged sheep currently used in the United States compare poorly with New Zealand, Irish, and British lines selected for efficient growth on pasture. (D. Flack)

GOATS

Goats can adapt to a wide range of conditions, but they have certain unique nutritional characteristics, which require that they be treated and fed differently from cows or sheep for best results. Despite the jokes about goats eating tin cans, their nutritional needs actually are higher than those of other ruminants. Milking goats will eat up to about 5.5 percent of their bodyweight equivalent in feed each day. In contrast, cattle and sheep usually eat 3 to 4 % of their bodyweight in feed per day.

Probably because of their high nutritional needs, goats are very selective grazers, and will use a pasture poorly unless managed well. Goats graze pasture forage from the top downwards in layers. Their level of dry matter intake and performance is very sensitive to pasture height and mass. Compared to young sheep, dry matter intake of young goats decreases at a faster rate as pasture mass declines. Goats appear to stop grazing at a pasture mass of about 900 lb DM/acre (about 1 inch tall), which certainly is close enough. A forage allowance of 15 lb DM/head/day appears to result in maximum productivity. Always have some hay available to goats grazing lush pasture, to satisfy their need for fiber. The best way to feed high-producing goats is to provide both high-quality pasture and access to good browse.

Goats can graze with cattle, and through this companion grazing produce more productive, higher quality pasture. Due to complimentary grazing habits, different preferences, and variation in forage, 1 to 2 goats can be grazed with every cow, and mutually benefit.

Does must have a dry period of at least 50 days before kidding. As soon as does are dried off, they can be maintained on pasture forage containing less energy than they were eating while milking. Feeding woody browse at this time is also helpful in restoring the function and capacity to their rumens and digestive systems, which may have changed a great deal and decreased in size during the months when they grazed only high-quality forage. During this time, does can be used to graze rough, brushy pastureland that needs improving. Be careful that does aren't stressed in using them to mob-stock rough pastureland.

Kids should be allowed to nurse their mothers for at least 4 days before being weaned to a bottle. This helps to condition the mothers' udders and gives the kids the best start possible. Make certain that does evenly milk out. (A virus transmitted through milk currently is a major health problem in goats. If it is present in the herd, all colostrum and milk should be heat-treated before feeding to kids.)

In dairy herds, kids can be given milk "replacer" at 4 to 7 days of age so that the mother's milk can be sold. When separated from their mothers for weaning, keep kids in small pens where they can be seen, smelled, and heard, but not nursed by the mothers for about 10 days. After this time they can run with their mothers without any problem of nursing the mothers. If you don't do something like this, you'll be driven crazy by the mothers who will devise any means to get you to substitute for their kids. Even if you can ignore the noise and commotion of the mothers, all of their activities require energy, and that energy could be put to better use in producing milk.

Kids should be offered forage when they are about 2

weeks old or less, to help develop their rumens and make weaning easier. Between 2 weeks of age and weaning, a kid needs 1 to 2 quarts of milk or milk substitute per day, plus pasture forage, water, salt, trace minerals, and sunshine. More than likely, similar equipment and methods used for rearing calves on pasture can be used for kids (see discussion in this chapter) Kids can be weaned at 6 to 8 weeks or as soon as they reach 2.5 times their birth weight. The earlier the weaning before they reach this size, the more their growth rate slows. To have no slowing of growth rate, weaning should be very gradual, and not begin until kids are at least 6 weeks old.

In herds kept for meat production, female kids needed for reproduction should be grazed with their mothers during as much of the milk feeding period as possible and not weaned early. In dairy herds, female kids grazing well-managed pastures reach liveweights of 70 pounds by the time they are about 7 months old; then they can be bred at about 8 months of age. Kids raised for slaughter should be weaned when they weigh 45 to 55 pounds, depending on the market. (A. Pell)

HORSES

Of all livestock, horses especially do well when grazed on good pasture. The freedom and exercise they experience under pasture conditions develop strong bones, muscle, and spirit.

The efficiency and years of service of horses are largely determined by how well their nutritional needs are met. Like other animals, horses require nutrients for maintenance, growth, reproduction, and production, which in horses is for working. Unlike other animals, the work of horses generally is very strenuous and irregular, which stresses them and makes it difficult to feed them according to their needs. Another complication is that the nutritional needs of horses may not remain the same even from day to day. Besides a horse's health, condition, and temperament, many factors influence the nutritional needs of horses. These include: age and size, stage of gestation or lactation, kind and amount of activity, weather, and quality and quantity of feed available. So besides feeding tables used as guidelines, a lot of experience, skill, and good judgment are needed to feed horses properly.

Well managed pastures, with salt and minerals available free-choice, can provide most or all of the nutrients needed by horses. A mineral mixture containing two parts calcium to one part phosphorus should be available if the pasture is mainly grass. If the pasture contains mostly legumes, the mineral mixture should be no more than one part calcium to one part phosphorus.

Horses' highly selective grazing promotes nutrient transfer from grazed to ungrazed areas; potassium especially can become deficient in grazed areas. The best way to prevent a pasture from deteriorating is to have cattle either with or following the horses to eat the forage that horses won't graze. Otherwise you'll probably need to mow, harrow, and fertilize at least once a season.

PIGS

High-quality pasture can satisfy much of the nutritional needs of pigs at all stages of gestation, lactation, growing, and finishing. Since 75 to 95 percent of pig feeds used in the United States are derived from grains, large savings in feed costs can be realized by grazing pigs on well-managed pasture. Cost of producing pigs on pasture is about $5 less per pig than in confinement. Often pigs can provide greater dollar returns from an acre of pasture than any other kind of livestock.

Pigs on pasture develop a surprisingly large capacity for using forage when the feeding of supplements is limited, so they're forced to graze. Unlike ruminants, however, the monogastric pig must rely mainly on feeds having readily digestible carbohydrates to meet energy needs. The more complex carbohydrates (e.g. cellulose and hemicellulose) contained in forages are broken down only by microbial fermentation. Because the pig doesn't have a rumen where such fermentation could occur, fibrous components of forage aren't used efficiently by pigs. For this reason, pigs usually can't be grown or fattened satisfactorily on pasture forage alone.

Pasturing of pigs has many real advantages over drylot feeding:

• Well-managed pasture can save large amounts of grain and protein supplements. With properly balanced rations feed savings have been found to be:
  - 10 to 20 % less grain and 30 to 50 % less protein supplement needed per 100 pounds of pork produced;
  - 500 to 1000 lb of grain and 300 to 500 lb of protein supplement saved per acre of pasture;
  - 50% lower feed costs for brood sows;
  - 3 lb less mineral needed per acre of pasture used.
• Pasturing decreases nutritional deficiencies in pigs, because of the higher quality proteins, vitamin content, unknown growth factors, minerals, and antibiotics that pastures provide in comparison to drylot feeds.
• Pasturing pigs reduces labor costs by about 30 %. This results from less effort and expense in sanitation, especially in manure handling, and less labor needed because the pigs harvest part of their own feed. No harvesting method is as efficient as allowing animals to graze their feed.
• Pastures conserve the maximum nutrient value of manure. At least 80 % of nutrients eaten are returned directly to pasture soils.
• Grazing sows on well-managed pasture during their pregestation period is especially important. The sows don't become overly fat and have greater reproductive efficiency. Pastured sows have 1.5 times more pigs born per litter than sows fed a similar ration in drylot.
• Because pasture forages contain large amounts of

protein, energy, vitamins, and minerals, they provide excellent feed for sows during gestation. This is especially true if the forage mixture contains 7 to 10 % of deep-rooted plants, such as chicory, dandelion, and yarrow, which have high mineral contents. Besides these nutrients, many forages contain reproductive factors needed for good litter size and survival of newborn pigs.

- Pastured pigs grow more vigorously and there are fewer runts. This is due to a combination of green succulent feed, proper nutritional balance, exercise, and better sanitary conditions with less incidence of diseases and parasites.
- Well-managed pasture also provides excellent feed for sows and promotes milk flow for their litters. Pigs nursed on pasture have a higher survival rate and are larger at weaning than pigs nursed in drylot.
- Weaned pigs on excellent pasture gain weight faster and reach market weight 1 to 2 weeks earlier than those fed in drylot. Contrary to opinion, the restricted exercise of animals fed in confinement doesn't result in more rapid and efficient liveweight gains.
- Dressing percentage is 72 % for pastured pigs, versus 64 % for those fed in confinement. This is very important for marketing because of the emphasis on lean meat.
- Pasturing pigs is perceived by the public and potential consumers as being a more humane way of raising them than in confinement. Hog producers need all the consumers they can get, and they don't need problems with people concerned about animal rights.
- Pastured pigs have a pleasant smell. Potential customers can be shown how their food is being raised. Can the same be stated about reeking confinement operations?

What all this means is that pigs can be raised on well-managed pasture with greater profitability and marketability than in drylot.

Pigs have a natural tendency to dig and root in soil. If

they have plenty of good-quality forage to eat, they tend to root less. But to use management intensive grazing properly, pasture forage must be uniformly grazed down to about 2 inches from the soil surface in each occupation period. Forcing pigs to graze so close to the ground may cause them to root up the soil. If they begin to dig, the pasture must be protected by putting rings in their noses, to keep them from rooting up the plants and generally destroying the pasture plant community.

High stocking densities of 60 to 80 animal units per acre (1 animal unit = 1000 lb liveweight) encourage pigs to graze uniformly. This requires careful management and close observation of pasture condition and frequent moves to fresh paddocks. Pasture masses should be about 2100 lb DM/acre pregrazing (4 inches tall) and 1200 to 1400 lb DM/acre postgrazing (2 inches tall).

Grazing other animal species (e.g. cattle, sheep, poultry) before, with, or after pigs helps to maintain a desirable pasture plant community and needed soil cover. Loss of soil cover and consequent soil erosion can be serious problems, especially in farrowing areas unless grazing is carefully planned. For example, Tom and Irene Frantzen of Iowa graze stocker cattle ahead of pigs to harvest surplus forage in early spring. Then they place farrowing sows in the pasture at a rate of four farrowing huts per acre. They arrange the huts so that traffic is controlled and damage to soil is limited. After the pigs are weaned and the sows removed, damaged areas can be reseeded and allowed to recover. Later in the season, the area is grazed again by stocker cattle. They have found that cattle will follow pigs only after at least a 30-day recovery period.

Farmers have found that Tamworth pigs perform better on pasture than other breeds. They are able to make good use of alternative feeds including pasture, alfalfa, rape, turnips, and food waste. Tamworth sows have good mothering ability for outdoor birthing and rearing.

Parasite levels must be kept low in pastures used by pigs. This can be accomplished if every 2 or 3 years a pasture is rested completely from grazing by pigs for 2 years, to allow parasites time to die in the absence of their host. During that time the pasture can be grazed by other livestock or machine-harvested.

Providing a free-choice mixture of diatamaceaous earth, kelp meal, and 10 pounds each of salt, rock phosphate, and limestone helps control parasites and meet mineral needs. Clean, cool water must be available at all times.

POULTRY
Poultry produced on pasture? Of course! That's the way it's done on small, diversified farms, and that's the way it used to be done on all farms before this latest trend toward confinement feeding of all livestock began. Confinement feeding of livestock was encouraged by the availability of inexpensive grains. Economically I suppose that made sense, but from the viewpoints of ecology, soil erosion, food quality, and responsibility for the world community, it never makes any sense to feed grain to livestock unnecessarily. Grain never really is cheap to produce in terms of energy used and costs of soil nutrients and erosion.

Chickens, turkeys, ducks, and geese can use pasture forage to the same advantage as other livestock, as long as the pasture and poultry are well managed. In many cases, the return per acre of good pasture used by poultry will be more than from its use by other livestock. The number of birds able to be carried per acre of pasture varies, depending on kind, quality, and quantity of forage available, weather, and soil texture and fertility.

Careful attention must be paid to grazing height, because chickens and other poultry can easily graze plants too low. Their grazing should be managed so that they don't graze lower than 1 inch from the soil surface.

Occupation periods must be short to prevent overgrazing and digging of the pasture.

Provide shade and shelter on the pasture so that poultry can get out of direct sunlight and rainfall when they want to, and have a place to lay eggs in and roost at night, protected from predators (e.g. cats, dogs, foxes, skunks, weasels, coyotes, raccoons).

Movable coops or cages can fill all of the above needs. Moving a cage following cattle through their pasture rotation is a good example of how poultry can be pastured (see Joel Salatin's *Pastured Poultry Profits* in References for a complete description of how to raise and market pastured poultry).

A 10- x 12-foot x 2-foot-high floorless cage can be used to pasture 75 to 100 poultry for meat. A 6- x 8-foot coop on bicycle wheels works well for small laying flocks, either confined to paddocks or allowed to range freely. Hexagon-shaped paddocks can be formed by two pairs of three 10- x 4-foot panels made from 3/8-inch iron rod and chicken wire. By rotating the paddock around the mesh-floored coop, several days' of fresh pasture can be gotten per move of the coop.

For larger flocks (e.g. 100 layers) a 12- x 20-foot coop on wheels can be moved with a tractor to pasture the free-ranging flock. The hens range up to about 200 yards away from the coop. Move the coop one paddock behind cattle so that the poultry can eat insects uncovered by the cattle, and fly larvae hatching every four days in fresh cattle poop. In the process, the poultry not only satisfy some of their protein needs from the insects they eat, but they break up and spread out the cow pies. This helps to decompose the manure and recycle its nutrients, decrease forage rejection around pies in subsequent rotations, and decrease fly populations.

Besides dramatic savings in feed costs, poultry grazing behind cattle can benefit overall pasture production. By eating insects, especially the developing fly larvae in cattle manure, poultry help reduce fly and other insect problems. By breaking down manure pies in their search for fly larvae, poultry enable nutrients contained in the manure to cycle faster, and zones of repugnance to disappear sooner, resulting in less rejected forage and more net forage production.

Chickens

Most people don't even know what good chicken or eggs taste like, because in their lifetimes they haven't eaten any that were produced on pasture. I know that the things produced in confinement are called chickens and eggs, but that's where the similarity ends. You can do it another way.

Actually, there are very good economic reasons for producing chickens and eggs on pasture:

- It may cost only half as much to feed chickens for meat and egg production on well-managed pasture, as it does in confinement.
- Money also can be saved in housing and management. By keeping hens on pasture with portable coops for laying and roosting until they finish laying in the fall, laying quarters in the main coop(s) can be made available for a new crop of spring pullets.
- Pullets also can be grazed on pasture, with large savings of feed possible. Pullets on pasture gain weight more rapidly, need less feed per pound of gain, and have better vigor than pullets raised in confinement.

Important considerations of well-managed poultry pasture, besides the plants available, are that pasture soils be free of poultry disease contamination, and that young birds graze well isolated from older birds to prevent spreading diseases. Rotating birds through the pasture area and periodic (every 2 years at least) resting of pasture areas completely from poultry grazing (machine-harvest the areas or graze with other livestock) for 2 or 3 years, should prevent soils from becoming contaminated with diseases.

All legumes (especially white clover), grasses, and forbs (e.g. plantain, dandelion, lambsquarters) commonly present in pastures provide excellent feed for chickens and other poultry. All required protein, vitamins, energy, and most minerals can be provided by well-managed pasture. Plants alone may not provide all needed amino acids, but grazing poultry also eat insects and worms, which supplement the plant diet. The only minerals that must be provided are sodium and chlorine, and possibly calcium and phosphorus. These can be provided easily by making available free-choice in separate containers common loose salt (sodium chloride), oyster shell or ground limestone, and defluorinated rock phosphate. Clean, cool, good quality water must always be available.

Chickens on pasture must be managed so that they use forage to the fullest extent possible. (For earliest maturity and highest initial level of egg production, feed for pullets on pasture should only be restricted by about 15 percent of what they would receive in confinement.) Any feed that's provided should be considered as a supplement to pasture forage; otherwise chickens will depend on the feed rather than forage. Supplements must be appropriate for the kind and quality of forage available. Chickens can be grown and eggs can be produced quite successfully on pasture with only supplements of minerals and small amounts of whole wheat or corn. (see Salatin, References)

Turkeys
All of the above applies equally to producing turkeys on pasture, except that turkeys must be completely isolated from chickens. This is mainly to prevent infection of the turkeys with black-head disease, which can be transmitted by infected carrier chickens. If chickens have grazed a pasture, allow a 2-year interval before using it for turkeys.

Ducks and Geese
Everything stated for producing chickens on pasture also

applies to ducks and geese, except they don't need any shelter from the rain; instead they need a plentiful source of clean water for bathing. Of all poultry, geese and Muscovy ducks are the best foragers and easiest to keep on pasture.

REFERENCES

Bartholomew, H. and F. Martz. 1995. Finishing cattle on pasture.*Stockman Grass Farmer.* 52(7):1.

Belanger, J. 1975. *Raising Milk Goats the ModernWay.* Garden Way Pub., Charlotte, Vermont.152 p.

Berry, W. 1981. *The Gift of Good Land.* North Point Press, San Francisco. 281 p.

Bliss, D.H. and G.H. Myers. 1996. Parasite control: strategies for dairy cattle in the 1990s. Hoechst-Roussel Agri-Vet Co., PO Box 2500, Somerville, NJ 08876-1258.

Blowey, R.W. 1985. *A Veterinary Book for Dairy Farmers.* Farming Press, Suffolk, England. 397 p.

Brickner, G. 1993. Seasonal breeding for dairy cattle. *Stockman Grass Farmer.* 50(12):14-5.

Calderwood, L. 1993. It's the water, stupid! *Udder Ideas.* University of Vermont Extension Service. December.

Campau, K. 1994. More on goats. *Stockman Grass Farmer.* 51(2):3-4.

Church, D.C. 1984. *Livestock Feeds and Feeding.* O & B Books, Corvallis, Oregon. 549 p.

Commoner, B. 1972. *The Closing Circle.* Alfred A. Knopf, New York.

Cramer, C. 1992. Hogs just migh be the ideal grazers. *The New Farm.* Sept./Oct. p.18-23.

Cromwell, G. 1984. Feeding swine. p. 389-412. In: D.C. Church (ed.) *Livestock Feeds and Feeding.* O & B Books, Corvallis, Oregon.

Ensminger, M.E. 1952. *Swine Husbandry*. Interstate Printers and Publishers, Danville, Illinois. 378 p.

Ensminger, M.E., and C.G. Olentine, Jr. 1978. *Feeds and Nutrition -- Complete*. Ensminger Pub. Co., Clovis, California. 1417 p.

Flanagan, J.P., and J.P Hanrahan. 1987. *Lowland Sheep Production-- Blindwell System*. An Foras Taluntais, Belclare, Tuam, Ireland. 20 p.

Foster, L. 1996. Irish pasture management. *Stockman Grass Farmer*. 53(10):6.

Frantzen, T. 1993. Pastured pigs not always sustainable. *Stockman Grass Farmer*. 50(5):16-17.

Green, J.T. 1993. Grazing habits and forages for goats. *Stockman Grass Farmer*. 50(6):8.

Guss, S.B. 1977. *Management and Diseases of Dairy Goats*. Dairy Goat Pub., Scottsdale, Arizona. 222 p.

Henderson, D.C., H.C. Whelden, Jr., and G.M. Wood. 1953. *Range and Confinement Rearing of Poultry*. University of Vermont Agricultural Experiment Station Bulletin 568.

Henning, A. 1994. How much grass and cash for the dry period? *Stockman Grass Farmer*. 51(10):13.

Hoechst-Roussel Agri-Vet Co. 1992. Effect of a herd health program. *Dairy Technical Bulletin*. Somervill, NJ. No. 1.

Hoechst-Roussel Agri-Vet Co. 1991. *Does your cattle dewormer control Nematodirus?* Somervill, NJ. 4 p.

Hoke, D.B. 1997. *A New Troubleshooter's Guide To Dairy Cows*. Northeast Organic Farming Association of Vermont, Bridge Street, Richmond, VT 05477. 75 p.

Honeyman, M. 1996. Swine production economics. *Stockman Grass Farmer*. 53(7):35.

Hughes, H.D. and M.E. Heath. 1967. Hays and pastures for horses. p. 671-683. In: H.D. Hughes, M.E. Heath, and D.S. Metcalfe (eds)

Forages. Iowa State University Press, Ames, Iowa.

Jackson, W. 1980. *New Roots for Agriculture.* Friends of the Earth, San Francisco. 155 p.

Jagusch, K.T. 1973. Livestock production from pasture. pp. 229-242. Tables 7-2 through 7-11 were adapted from p. 234-241. In: R.H.M. Langer (ed.) *Pastures and Pasture Plants.* A.H. and A.W. Reed, Wellington, New Zealand.

Jones, V. 1994. Selenium issue is two-pronged. *Stockman Grass Farmer,* 51(8):29-30.

Jones, V. 1996. Animal health. *Stockman Grass Farmer,* 53(4):25-26.

Kennard, D.C. 1967. Forage for poultry. p. 663-670. In: H.D. Hughes, M.E. Heath, and D.S. Metcalfe (eds) *Forages.* Iowa State University Press, Ames, Iowa.

Kidd, R. 1993. Control parasites organically? *The New Farm.* Nov./Dec. p. 7-11.

Kruesi, W.K. 1985. *The Sheep Raiser's Manual.* Williamson Pub., Charlotte, Vermont. 288 p.

Kruesi, B. 1992. Lambing on pasture can reduce capital expenses. *Stockman Grass Farmer.* 49(4):14-16.

Kunkel, J.R., W.M. Murphy, D. Rogers, and D.T. Dugdale, Jr. 1983. Seasonal control of gastrointestinal parasites among dairy heifers using two strategically timed treatments of fenbendazole. Bovine Practitioner. 18:54-57.

Landa, P. 1993. Grass finishing in Argentina. *Stockman Grass Farmer.* 50(10):9-10.

Mackenzie, D. 1957. *Goat Husbandry.* Faber and Faber Ltd., London.

Mayne, C.S. 1991. Effects of supplementation on the performance of both growing and lactating cattle at pasture. British Grassland Society Occasional Symposium No. 25, p. 55-71.

Mayne, C.S. 1992. Supplemental feeding found to be of little value.

The Stockman Grass Farmer. 49(7):1,7.

Mayne, C.S. 1996. USA lacks grass confidence. *Stockman Grass Farmer.* 53(2):10.

McCall, D.G., and M.G. Lambert. 1987. Pasture feeding of goats. p. 105-109. In A.M. Nicol (ed..) *Feeding Livestock on Pasture.* New Zealand Society of Animal Production, Hamilton, New Zealand.

McCarville, P.J. 1992. *McCarville's Calf Rearing System.*   18875 McCarville La., Mineral Point, WI 53565. 2 p.

Michel, J.F. 1974. Arrested development of nematodes and some related phenomenons. Advances in Parasitology. 12:279-366.

Morand-Fehr, P. and D. Sauvant. 1984. Feeding goats. p. 372-388. In: D.C. Church (ed.) *Livestock Feeds and Feeding.* O & B Books, Corvallis, Oregon.

Mott, G.O., and C.E. Barnhart. 1967. Forage utilization by swine. p. 655-662. In: H.D. Hughes, M.E. Heath, and D.S. Metcalfe (eds) *Forages.* Iowa State University Press, Ames, Iowa.

Nation, A. 1992. Productivity standards needed for graziers. *Stockman Grass Farmer.* 49(2):1, 6-7.

Nation, A. 1992. New Zealand style calf rearing catching on in Wisconsin. *Stockman Grass Farmer.* 49(9):9-10.

Nation, A. 1992. Early weaning pays with replacements. Allan's Observations. *Stockman Grass Farmer.* 49(11):30.

Nation, A. 1992. Pennsylvania dairymen are finding...the cows know best. *Stockman Grass Farmer.* 49(12):15-17.

Nation, A. 1992. Paradigm shift. *Stockman Grass Farmer.* 49(12):21.

Nation, A. 1992. *Pasture Profits with Stocker Cattle.*   The Stockman Grass Farmer, P.O. Box 9607, Jackson, MS 39286-9909. Phone: 800-748-9808. 192 p.

Nation, A. 1993. Production per man more important than per cow. *Stockman Grass Farmer.* 50(3):1,6-10.

Nation, A. 1993. Organic hogs on pasture. *Stockman Grass Farmer.* 50(4):20.

Nation, A. 1994. Kiwis find USDA choice is possible from grass. *Stockman Grass Farmer.* 51(7):1,9-10.

Nation, A. 1994. Horses grazed alone soon make pastures sick. *Stockman Grass Farmer.* 51(8):41.

Nation, A. 1994. Dairy replacements more profitable than beef stockers. *Stockman Grass Farmer.* 51(10):6, 9.

Nation, A. 1994. Whole milk on pasture rearing systems offer lower costs and healthier calves. *Stockman Grass Farmer.* 51(11):7.

Nation, A. 1995. The problem of feed substitution. *Stockman Grass Farmer.* 52(4):26.

Nation, A. 1996. Vermont research shows 85/15 paddock split could pay you an extra $1000/month. *Stockman Grassfarmer.* 53(10):13.

Nation, A. 1997. Grass fat is good for you. *Stockman Grass Farmer.* 54(7):9.

Nicol, A.M., D.P. Poppi, M.R. Alam, and H.A. Collins. 1987. Dietary differences between goats and sheep. Proceedings of New Zealand Grassland Association. 48:199-205.

Poincelot, R.P. 1986. *Toward a More Sustainable Agriculture.* AVI Publishing Co. 241 p.

Poppi, D.P., T.P. Hughes, and P.J.L'Huillier. 1987. Intake of pasture by grazing ruminants. p. 55-63. In A.M. Nicol (ed..) *Feeding Livestock on Pasture.* New Zealand Society of Animal Production, Hamilton, NZ.

Salatin, J. 1991. Pastured egg production. *Stockman Grass Farmer.* 48(3):1,4.

Salatin, J. 1991. Profit by appointment only. *The New Farm.* Sept./Oct. p. 8-12.

Salatin, J. 1993. *Pastured Poultry Profit$.* Polyface, Inc., Rt. 1, Box 281, Swoope, VA 24479. Phone: 703-885-3590. $30. 340 p. (Video also available for $50)

Salatin, J. 1996. Weaning the low stress way. *Stockman Grass Farmer.* 53(6)24.

Sampson, R.N. 1981. *Farmland or Wasteland: a Time to Choose.* Rodale Press, Emmaus, Pennsylvania. 422 p.

Schafer, D. 1993. Weaning ourselves (about weaning beef calves). *Stockman Grass Farmer.* 50(10):15.

Shirley, C. 1993. Pasture pigs of lean pork, low overhead. *The New Farm.* Nov./Dec. p. 20--24, 58.

Simmons, P. 1976. *Raising Sheep the Modern Way.* Storey Communications, Pownal, Vermont. 234 p.

Smith, B. 1993. Reproductive cycles. *Stockman Grass Farmer.* 50(4):11-12.

Stone, S. 1995. Genetics for pastured pigs. *Stockman Grass Farmer.* 51(12):3, 20.

Subcommittee on Beef Cattle Nutrition. 1984. *Nutrient Requirements of Beef Cattle.* National Academy of Sciences, Washington, D.C. 6th ed.. 55 p.

Tranel, L.F. 1993. Supplementing grain on pastures. *Stockman Grass Farmer.* 50(11):17.

Van Soest, P.J. 1982. *Nutritional Ecology of the Ruminant.* O & B Books, Corvallis, Oregon. 374 p.

Voisin, A. 1959. *Grass Productivity.* Island Press, Washington, D.C.

Weiss, B. 1991. Pasture-based feeding systems. Presentation at Mahoning Co. Farm Field Day. Ohio Agricultural Research and Development Center, Wooster.

Welch, J.G. 1993. Concentrate feeding levels for lactating cows on pasture. In W.M. Murphy (Project Coordinator) USDA SARE Program. Final Report. Dept. Plant & Soil Sci., University of Vermont.

8
Feed Planning

A cold rain starting
And me without a hat.
On second thought, who cares?

Basho

Achieving any goal requires planning. Farmers using Voisin management intensive grazing know that the unpredictable nature of pasture growth causes problems in planning the feeding of their livestock.

When plant growth is less than livestock needs, total pasture supply on the farm decreases and eventually the animals could run out of feed. Then either supplemental forage or more pasture must be provided to maintain production.

When pasture plants grow faster than livestock can eat them, a surplus accumulates. Unless the surplus is controlled by conserving the forage (machine harvesting and storing) or grazing more animals, forage quality and sward density decrease and plant growth slows.

If you can anticipate when such situations are likely to occur, you can use appropriate management (e.g. change stocking rate, calving and dry-off dates, conserve forage surplus) to deal with them. Preparing a feed plan can help you achieve optimum annual feed use. Three kinds of

---

This chapter was adapted mainly from Milligan, K.E., I.M. Brookes, and K.F. Thompson. 1987. Feed planning on pasture. p. 75-88. In. A.M. Nicol (ed.) *Feeding Livestock on Pasture.* New Zealand Society of Animal Production. Hamilton. Occasional Publication No. 10. During 1995 we were fortunate to have Dr. Ian Brookes with us in Vermont. He helped us to better understand feed planning and to simplify this chapter so others might understand it as well. Ian provided the Feed Plan Worksheet and example of planning for a pasture-based dairy.

feed plans are used, depending on the decisions that need to be made and the period that they cover: feed profile, feed budget, and grazing plan.

Phil Taylor, a New Zealand farmer, explained it this way: "Being a successful pasture farmer is easy. All you have to do is harvest all the grass you grow, never have a surplus, and never run out!"

What follows may seem complicated at first, but it all boils down to: **1.** How much grows? **2.** How much is eaten? **3.** How much accumulates?

A feed plan essentially is similar to a cash budget or bank account:

Starting balance + income - expenses = ending balance

Starting pasture cover + plant growth - forage eaten and conserved = ending pasture cover

FEED PROFILE

A feed profile shows the average or most likely pattern of feed supply and demand over a year. Once you estimate feed supply and demand, you can develop a feed profile to decide on potential stocking rate and to balance seasonal feed demand with what you expect the pattern of pasture forage supply to be.

A feed profile indicates when and about how much surplus forage can be conserved or stockpiled and carried forward as increased pasture cover (average pasture mass). It can also provide an estimate of the amount of supplemental feed that may be needed. Calving and lambing can be timed to synchronize increased feed demand of lactation with high rates of pasture growth.

FEED BUDGET

A feed budget helps you to make management adjustments during the season. It provides information on how best to use available feed to achieve optimum animal production levels and the most profitable use of pasture forage. Since pasture supply and demand usually don't balance the first time that you calculate a feed budget, you have to decide on the least costly way of overcoming a forage deficit, or the most profitable way of using surplus forage. When you get the budget to balance, then you can prepare a grazing plan to ration pasture forage to achieve planned dry matter intakes.

Feed budgets can be calculated in different ways, depending on your situation. This is a simple way:

1. Calculate pasture forage supply:

Pasture forage supply (lb DM/acre) =
   present pasture cover (lb DM/acre) +
   pasture plant growth (lb DM/acre/day x days) -
   pasture cover at end of season (lb DM/acre)

2. Calculate feed demand:

Feed demand (lb DM/acre) =
   stocking rate (number of animals/acre/season) x
   daily dry matter intake (lb DM/head) x
   number of days in grazing season

Then compare supply with demand, and calculate the surplus, deficit, or balance. If a deficit occurs, first check the estimated plant growth rates and final pasture cover to make certain that they are correct. If there's still a deficit, you have to evaluate the cost of supplemental feeds, compared to a lower production level from reduced feed intake. If there's surplus forage, you need to consider the effects of increasing feed intake, high postgrazing residues, or of using the excess in some other way, such as conserving it or bringing in more animals.

You should regularly monitor pasture cover to know if it becomes different from what you predicted in your feed budget. If cover does change, you can make the appropriate management adjustment in time.

GRAZING PLAN

A grazing plan is the order paddocks will be grazed during the next 7 days or more. It helps decide how long livestock will graze each paddock to eat the forage they need and leave the paddock in good condition for the next grazing. Before preparing a grazing plan, your feed budget should balance, indicating that there will be enough pasture available, or if and when supplements will be needed.

To make a grazing plan, you have to know the pasture mass or cover and area of every paddock. So you make the grazing plan after your weekly walk over the pasture and you have estimated pasture mass in all paddocks and averaged it for the farm. Then list paddocks in the order they'll be grazed, from highest to lowest pasture mass. Also take into consideration location of paddocks within your farm and ease of livestock movement among the paddocks. Then calculate the number of days that animals will occupy each paddock.

--∞--

When making feed plans, you must consider subsequent effects that your management will have on plant and animal production. For example, swards maintained very short and with low pasture mass will have a high content of green leaves, but slower plant growth rate. In contrast, laxly grazed swards with high residual can result in a lot of rejected forage, reduced forage feeding value, decreased sward density, and eventual decreased productivity.

FEED SUPPLY

Pasture plant growth rate, pasture cover, and supplements make up the feed supply. The contribution of each of these parts differs depending on your goal and planning.

Pasture Growth Rate

The time between successive grazings is determined by the rate of pasture plant growth. For example, if a paddock grazed to a residual of 1350 lb DM/acre needs to recover to

2400 lb DM/acre before being grazed again, this will take 21 days if pasture grows at 50 lb DM/acre/day. If daily growth drops to 25 lb DM/acre, then recovery time will be 42 days (2400-1350=1050÷50=21 or 1050÷25=42).

You can estimate pasture growth rate by regularly assessing changes in pasture cover on paddocks not grazed over a period of several days. For example, if pasture mass in a paddock increases from 1350 to 1700 lb DM/acre in 7 days, the average growth rate is 50 lb DM/acre/day (1700-1350=350÷7=50). With experience you can visually estimate pasture mass to within 100 to 300 lb DM/acre of the actual amount, or you may prefer to use a measuring device such as a rising plate meter or pasture probe.

Very little information exists about average pasture growth rates under management intensive grazing in the United States. The data for locations in Vermont and West Virginia presented in Table 8-1 and Figure 8-1 are examples of information needed to make feed profiles. If you keep records of your estimates over several years, they can provide a good indication of the likely seasonal range of forage production in individual paddocks and the whole pasture.

Table 8-1. Average plant growth rates of Kentucky bluegrass-orchardgrass-white clover permanent pastures under management intensive grazing (1989-1990)

Month	Vermont	West Virginia
	----lb DM/acre/day----	
April	33	53
May	59	64
June	57	70
July	42	54
August	40	36
September	33	26
October	12	18

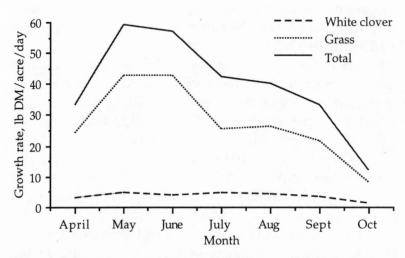

Figure 8-1. Seasonal pattern of grass (Kentucky bluegrass, orchardgrass) white clover, and total net forage production in pasture under management intensive grazing near Colchester, Vermont, during 1989-1990.

Pasture Cover

Pasture cover refers to average pasture mass (lb DM/acre) on the farm at a certain time. Calculate pasture cover by multiplying pasture mass in each paddock by the paddock area (acres) to estimate total forage in the paddock. Add the amounts of forage in individual paddocks together, then divide by total pasture area, to get pasture cover (Table 8-2).

Suitable pasture cover on any farm depends on such things as expected plant growth rate, stocking rate, and time of birthing. Usually you should try to maintain pasture cover above 900 lb DM/acre, and less than 2200 lb DM/acre. The pasture cover of your feed plan determines if the forage allowance needed for planned dry matter intake can be achieved.

Table 8-2. Example of pasture cover calculation.

Paddock number	Pasture mass per paddock	Paddock area	Total forage
	lb DM/acre	acre	lb DM
1	1650	1.0	1650
2	2000	1.5	3000
3	2200	1.6	3520
4	1250	1.9	2375
5	1850	1.0	1850
6	2550	1.0	2550
7	1500	1.5	2250
8	2400	2.0	4800
9	1400	1.0	1400
10	1200	2.0	2400
	Total	14.5	25795

Pasture cover = 25795 ÷ 14.5 = 1779 lb DM/acre.

Changes in pasture cover over time express the relationship between supply and demand. If you regularly monitor pasture cover, you can determine if problems likely will occur. You can then develop a plan of action to prevent adverse effects on production and achieve your goal of optimum livestock production and profitability from pasture.

A good way to keep track of the whole schmeer is to graph the pasture cover expected at the end of each month (Figure 8-2); then you can add the actual cover values that you estimate as the season progresses. If the actual value is much lower than the expected one, there might be a feed deficit that may not be obvious yet. This can provide a warning of a future problem that is developing, before it becomes serious, and enable you to take action to avoid it.

Too low or high pasture cover indicates that you need to do something. Pasture cover can be increased by:

• Changing the timing of events such as birthing to

achieve a better balance between supply and demand.
- Increasing forage allowance by making more pasture area available.
- Feeding supplements such as hay, green chop, or haylage.

High pasture cover (greater than 2200 lb DM/acre) indicates that:
- You need to set aside areas to conserve surplus forage,
- Bring in more livestock to increase the demand, or
- Change the time of birthing.

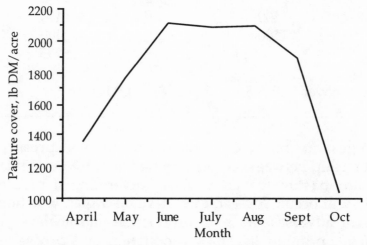

Figure 8-2. Example of expected pasture cover (values from Table 8-2).

Supplements
Supplemental feeds such as haylage, green chop, hay, grain, and concentrates form part of the total feed supply. Supplements substitute for pasture forage, so be careful in feeding them because they can reduce pasture forage intake and increase your costs unnecessarily.

Pasture Surplus
Conserving excess forage as hay or haylage reduces pasture surpluses that occur when plant growth rates are

greater than feed needs. It decreases pasture cover in areas that are harvested, and helps maintain the overall pasture in good condition for subsequent grazings. It is also a way of shifting surpluses to times of shortage. Conserved surplus forage can be used to reduce pasture deficits during the grazing season, fed out in winter, or sold to others to meet their feed deficits.

FEED DEMAND

Feed demand is the total amount of dry matter that will be needed during a certain period. Pasture dry matter need is calculated from the average daily dry matter intake needed per animal, multiplied by the number of animals and number of grazing days.

Forage Quality: Protein And Energy

Feed demand is calculated based on the energy needed to meet production goals such as milk production level, body condition score, liveweight gain, total liveweight, and birthing date.

In well-managed North American pastures, forage protein levels are more than adequate for all livestock production. Protein levels actually are too high for the amount of energy available in the forage. The excess protein has to be excreted by grazing animals, unless energy supplements are fed to make use of the protein. For this reason feeding energy supplements may increase efficiency of pasture feeding. Of course, profitability of feeding energy supplements depends on their cost and value of the production increase obtained.

Since protein content is usually adequate for animals grazing well-managed pastures, there's no point in continuing to be concerned about it. For most practical situations, pasture forage can be evaluated and classified in terms of its energy content (Table 8-5). This enables easy use of forage energy values for feed planning under normal, changing feeding conditions.

To make this as simple and as useful as possible, this book shows NEL for dairy cows, and TDN and ME for all other livestock. You can use the energy value that you're most familiar with for evaluating your pasture and incorporating it into the total ration that you're feeding.

If you know the energy needs of livestock at different liveweights, production levels, and physiological states, and the energy value of the pasture forage, you can estimate:

• How much livestock product the pasture will yield.
• What the dry matter intake per animal should be.
• Forage allowance.
• How much pasture area is needed.
• The stocking rate or carrying capacity of your pastures.

Since you are accustomed to production levels of traditionally mismanaged pastures, the examples of production potentials possible from well-managed pastures shown in the dairy feed plan example and tables at the end of this chapter may seem to be fantastic, but these are real possibilities.

Of course, these production levels require excellent management and well-developed pasture. You probably won't achieve such high levels of production the first year that you switch from year-round confinement feeding or continuous grazing. Your pastures and soils will need time to repair the damage done from a century or more of abuse. You will need time to develop expertise in Voisin management intensive grazing, and that will only come with experience; it is an art. Your livestock will need time to adjust to the new situation, especially if they've been fed in confinement. Animals that have been bred for years to use large amounts of supplements may need a period of adjustment or selection to achieve more efficient use of protein and energy from pasture forage.

But you can begin now to move toward the production levels shown in the tables. And you might be pleasantly surprised to find that your pasture is not in such terrible shape as you thought. Luckily, nature is very forgiving, otherwise humans long ago would have destroyed their life-support system!

You can take encouragement from the results we got in 1984, when we collected and analyzed 497 pasture forage samples during a 5-month grazing season (May 1-October 1) on six dairy farms in northern Vermont (Table 8-3). For one of the farms it was its first season using Voisin management intensive grazing; two were in their second season, two were in their third season, and one was in its fourth season. Even though the farmers had been managing grazing for such a short time, average forage energy contents on the farms were all within the "good" class during the entire season (Table 8-5). Also, these values were for whole-plant samples. If milking or gaining animals graze paddocks first and get the best forage available, energy levels of that forage probably will be higher than in whole-plant samples.

TABLE 8-3. Analyses (dry weight basis) of forage from permanent pastures under Voisin management intensive grazing on 6 Vermont dairy farms, May 1-October 1, 1984.

									Analyses		
Month	DM	CP	AP	ADF	Ca	P	K	Mg	TDN	ME	NEL
						---%---					Mcal/lb
May	22	22	21	28	1.2	.5	1.5	.2	70.4	1.16	.73
June	22	20	19	30	1.0	.4	1.7	.2	67.8	1.11	.70
July	25	21	20	30	1.1	.4	1.7	.3	66.9	1.10	.69
August	24	23	22	28	1.3	.5	1.6	.2	70.4	1.16	.73
Sept.	20	25	23	26	1.4	.6	1.4	.2	71.3	1.17	.74
Avg	23	22	21	28	1.2	.5	1.6	.2	69.4	1.14	.72

Average yield = 3.7 tons DM/acre.

DM = dry matter, CP = crude protein, AP = available protein, ADF = acid detergent fiber, Ca = calcium, P = phosphorus, K = potassium, Mg = magnesium, TDN = total digestible nutrients, ME = metabolizable energy, NEL = net energy lactation.

During 1988 to 1990, 746 forage samples were collected and analyzed from well-managed pastures on farms in Maine, New Hampshire, New York, and Vermont as part of the USDA Low Input Sustainable Agriculture program. All of these analyses (Table 8-4) confirmed the high protein and energy levels possible in well-managed pasture forage that we measured in 1984. For comparison in Table 8-4, I separated the results for our farm. We collected forage samples every 2 weeks. During the 1988-1990 we had been using Voisin management intensive grazing of cattle and sheep on our farm for 9 years.

TABLE 8-4. Analyses (dry weight basis) of forage from permanent pastures under Voisin management intensive grazing in Vermont and the Northeast.

Analysis		Month						
		April	May	June	July	Aug	Sept	Oct
Vermont:								
6 farms, 1984 (dry forage yield = 3.7 ton/acre):								
DM, %			22	22	25	24	20	
CP, %			22	20	21	23	25	
TDN, %			70.4	67.8	66.9	70.4	71.3	
ME, Mcal/lb DM			1.16	1.11	1.10	1.16	1.17	
NEL, Mcal/lb DM			.73	.70	.69	.73	.74	
Murphy farm, 1988-90 (dry forage yield = 4.2 ton/acre):								
DM, %		16	19	18	18	18	15	17
CP,%		31	23	25	23	27	30	31
NEL, Mcal/lb DM	.75	.70	.68	.68	.74	.73	.80	
Northeast, 1988-90:								
DM, %			19	21	24	22	21	20
CP, %			24	20	22	21	23	24
NEL, Mcal/lb DM			.75	.66	.68	.69	.71	.73

DM = dry matter, CP = crude protein, TDN = total digestible nutrients, ME = metabolizable energy, NEL = net energy lactation.

--∞--

Analyze your pasture forage and use the energy values in management and planning. As you accumulate information and experience, each year your planning will get better, management will improve, production costs will decrease, and your farm's profitability will increase.

Feed Demand Calculations

Feed demand is calculated for groups of animals. Since there's a lot of variation among animals, calculate for the average production goal (e.g. milk production level) and use demand for the average animal.

The daily energy requirement for the average animal of a group to meet a desired production goal (see Tables at end of this chapter) must be converted to pounds of pasture dry matter intake required per day.

<u>Dry Matter Intake</u>

From the animal standpoint the amount of dry matter eaten per day depends on a lot of things, including number of bites per minute of grazing time, amount of forage eaten per bite, and total grazing time. Grazing management can influence all of these factors. The main way we can increase dry matter intake is to provide animals with a sward that is in a good physical (pasture mass: 1800-2400 lb DM/acre; short: 4-6 inches tall) and nutritional condition (young vegetative), so that they can graze easily and get a lot of high quality feed per bite.

Dry matter intake seems to be influenced mainly by daily forage allowance. Supplemental feed given to animals decreases their dry matter intake from pasture and increases residual, since animals fed supplements are less willing to graze closely.

Given the above complications and animal, sward, and sampling variability, it's no wonder that dry matter intake is the most difficult aspect to determine in pastured livestock. Any estimate of dry matter intake is just that: an estimate. Don't expect it to be precise. Average dry matter intake (lb DM/head/day) required from pasture can be estimated in at least two ways:

1) Divide average energy required for a production level by energy content of the pasture forage.

2) If pasture is well managed and forage is good to excellent quality, estimate 3 to 4.5% of average bodyweight for dairy cows (depending on supplementation level), 3% for stocker cattle, 4% for lactating sheep, 4.5% for growing lambs, and 5.5% for lactating goats and growing kids. These will be close to the amounts of dry matter that animals will eat each day.

FEED PLANNING IN PRACTICE

Feed plans are based on your predictions of future feed supply and demand. You must regularly monitor plant and animal production by walking and observing your entire pasture at least every 7 days, and adjust your plan and grazing management accordingly. One of the most variable parts of a feed plan is plant growth rate. If you monitor soil temperature and moisture, this information can help you predict plant growth and accumulation.

You can make feed plans based on predictions of high, low, or average plant growth rate, with management options planned for each level. Such exercises can help you determine when to make important decisions such as feeding supplements or selling livestock, so that the decisions can be made before reaching a critical point.

You can monitor your feed plan by comparing the actual amount of pasture cover, determined by your weekly pasture walk, with what you predicted. If more detail is needed, check plant growth rates, occupation periods per paddock, and pre- and postgrazing pasture masses every time you move animals during a rotation. If you regularly check milk production levels or weigh animals, you can determine if animal performance objectives are being reached.

FEED PLAN WORKSHEET

Pasture Demand

<u>Needed information</u>
Livestock numbers:
    Lactating   ____
    Dry ____
    Growing ____
Average liveweight, lb ____
Average production level, lb milk or gain/day ____
Average daily grain supplement level, lb ____
    and its energy content, NEL, TDN, or ME/lb DM ____

<u>Calculate energy and dry matter required from pasture</u>
Use the tables at the end of this chapter for estimates of daily energy required for maintenance and production of grazing livestock. Subtract the energy of any grain or other supplement fed from the average animal's total energy requirement, to obtain the energy needed from pasture. Then calculate the daily dry matter required per animal from pasture, using its energy content determined by forage analysis or an estimate:

Daily energy requirement per average animal ____
Less energy from average grain fed per animal ____
Energy required from pasture per day per animal ____
Pasture energy content, NEL, TDN, or ME/lb DM ____
Pasture demand (energy required÷pasture energy content),
    lb DM/animal/day ____
Total pasture demand ____

Pasture Supply
<u>Estimate pasture growth and cover on per-acre basis</u>

Walk your pasture every 7 days, estimate pasture mass in all paddocks (use eye estimate or measuring tool), and calculate how much pasture mass or cover has increased in ungrazed paddocks and divide by the number of days between measurements. (Cover in grazed paddocks will have decreased over the period.) For example:

| Paddock# | Measurements | | Daily growth |
	June 1	June 7	
	---------------lb DM/acre-------------		
1	1400	1775	54
2	2200	2500	43
3	2500	2800	43
4	2000	2350	50
5	2050	2350	43
6	1450	1800	50
7	1300	1700	57
8	1350	1800	64
9	1500	1800	43
10	1400	1800	57
		Average	50 lb DM/acre

If you rank paddocks according to their pasture mass from high to low after your weekly pasture walk, this becomes your grazing plan of the order that paddocks will be grazed during the next week.

Feed Budget

(a) Pasture area, acres _____

(b) Livestock number _____

(c) Pasture demand, lb DM/animal/day _____

Grazing season start and end dates _____
Cover at start of season, lb DM/acre _____
Pasture growth, lb DM/acre/day: enter in feed budget table

Construct a monthly feed budget table for the whole grazing season in lb DM/acre. Average the pasture DM requirements across all grazing acres of the farm.

Monthly Feed Budget Table

Month	Days	Pasture Grown	Eaten	Surplus	Period total	Cover
		-------lb DM/acre/day-------			---lb DM/acre---	
	(d)	(e)	(f=bc/a)*	(g=e-f)	(h=gd)	(start+h)

Feed Profile

1. Answer these questions:
• How many days grazing are available from start to end of grazing? Total days grazing (d): _____
• Total pasture grown? Days (d) x pasture grown (e) totaled over all months: _____ lb DM/acre.
• Total pasture eaten? Days (d) x pasture eaten (f) totaled over all months: _____ lb DM/acre.
• What surplus if any will accumulate over the grazing season? Total grown - total eaten: _____ lb DM/acre.
• In which months will there be feed surpluses and in which months will there be deficits?
Surpluses: _____
Deficits: _____

2. For each month calculate:
• How many days regrowth are needed until there is enough pasture mass to regraze a paddock? Divide total daily pasture demand by daily pasture grown: _____ days.
• How many acres are needed to grow the monthly pasture requirements of the milkers? Divide total pasture demand for the month by the total pasture grown during that period: _____ acres.
• How many acres could be left ungrazed?
• How much surplus dry matter will accumulate on these ungrazed paddocks? Multiply number of days by pasture

grown per day by number of ungrazed acres: _____ lb DM.

3. Consider the options throughout the period:
• How much area will you offer lactating or growing animals each month?
• Might you consider feeding conserved forage to the producing animals? If so, how much and when?
• What feeding plan will you use for the dry animals?
• How much forage will you conserve and when?

FEED PLAN EXAMPLE: PASTURE-BASED DAIRY

Pasture Demand

Livestock
50 milking cows, average 1300 lb liveweight, produce 65 lb milk/day, fed 15 lb grain (90% DM, 0.9 Mcal NEL/lb DM).

Energy and dry matter required from pasture
Use Table 8-6 for daily NEL requirement of milkers grazing good pasture. Subtract NEL of grain supplement from the average milkers' total NEL requirement to obtain NEL required from pasture; then calculate the daily pasture dry matter requirement per cow.

Grain NEL = lb x % DM x Mcal/lb DM
15 lb x 0.9 x 0.9 Mcal/lb DM = 12.2 Mcal/lb DM

	Milkers
Daily NEL required, Mcal	39.8
Less grain NEL, Mcal/lb DM	12.2
NEL required from pasture	27.6
Pasture NEL, Mcal/lb DM	0.75
Pasture demand, lb DM/cow/day	36.8
(NEL required÷pasture NEL content)	
Pasture demand for 50 milkers: 50 x 36.8 = 1840 lb DM/day	

Pasture Supply

Pasture area: 50 acres
Grazing season: April 15 - October 31
Cover at start of grazing: 2000 lb DM/acre
Daily pasture growth, lb DM/acre:

April	May	June	July	August	September	October
30	60	50	35	45	40	30

Feed Budget

Construct a monthly feed budget table in lb DM/acre. Average the pasture dry matter requirements across all 50 acres.

a) 50 acres of pasture
b) 50 cows
c) Pasture demand: 36.8 lb DM/cow/day; total demand for 50 cows: 1840 lb DM/day
Grazing season: April 15 - October 15
Cover at start: 2000 lb DM/acre

Monthly Feed Budget Table

Month	Days	Pasture			Period	Cover
		Grown	Eaten	Surplus	total	
		-------lb DM/acre/day-------			---lb DM/acre---	
	(d)	(e)	(f=bc/a)*	(g=e-f)	(h=gd)	(start+h)
April	15	30	36.8	-6.8	-102	1898
May	31	70	36.8	+33.2	+1029	2927
June	30	55	36.8	+18.2	+546	3473
July	31	30	36.8	-6.8	-211	3262
August	31	45	36.8	+8.2	+254	3516
Sept	30	25	36.8	-11.8	-354	3162
Oct	15	20	36.8	-16.8	-252	2910

*Pasture eaten = (number of livestock x pasture demand/animal) ÷ acres of pasture.

2. Answer these questions:

• How many days grazing are available from start to end of grazing? Total days grazing (d): 183.

• Total pasture grown? Days (d) x pasture grown (e) totaled over all months: 7645 lb DM/acre.
• Total pasture eaten? Days (d) x pasture eaten (f) totaled over all months: 6735 lb DM/acre.
• What surplus if any will accumulate over the grazing season? Total grown - total eaten: 910 lb DM/acre.
• In which months will there be feed surpluses and in which months will there be deficits?
Surpluses: May, June, August
Deficits: April, July, September, October

3. For each month calculate:
• How many days regrowth are needed until there is enough pasture mass to regraze a paddock. Divide total daily pasture demand by pasture grown per day:
April: 1840 lb DM/day ÷ 30 lb DM/day = 61 days;
May: 1840 lb DM/day ÷ 70 lb DM/day = 26 days
June: 1840 lb DM/day ÷ 55 lb DM/day = 33 days
July: 1840 lb DM/day ÷ 30 lb DM/day = 61 days
August: 1840 lb DM/day ÷ 45 lb DM/day = 41 days
September: 1840 lb DM/day ÷ 25 lb DM/day = 74 days
October: 1840 lb DM/day ÷ 20 lb DM/day = 92 days

• How many acres are needed to grow the monthly pasture requirements of the milkers? Divide total pasture demand for the month by the total pasture grown during that period:
April: (15 days x 1840 lb DM/day) ÷ (15 days x 30 lb DM/acre/day) = 61 acres
May: (31 days x 1840 lb DM/day) ÷ (31 days x 70 lb DM/acre/day) = 26 acres
June: (30 days x 1840 lb DM/day) ÷ (30 days x 55 lb DM/acre/day) = 33 acres
July: (31 days x 1840 lb DM/day) ÷ (31 days x 30 lb DM/acre/day) = 61 acres
August: (31 days x 1840 lb DM/day) ÷ (31 days x 45 lb DM/acre/day) = 41 acres

September: (30 days x 1840 lb DM/day) ÷ (30 days x 25 lb DM/acre/day) = 74 acres

October: (15 days x 1840 lb DM/day) ÷ (15 days x 20 lb DM/acre/day) = 92 acres

- How many acres could be left ungrazed?

April: none, plus additional grazeable land (e.g. hayland) or supplemental feed is needed

May: 24 acres

June: 17 acres

July: none, plus additional grazeable land or supplemental feed is needed

August: 9 acres

September and October: none, plus additional grazeable land or supplemental feed is needed

- How much surplus dry matter will accumulate on these ungrazed paddocks? Multiply number of days by pasture grown per day by number of ungrazed acres:

April: 0

May: 31 days x 70 lb DM/acre/day x 24 acres = 52080 lb DM

June: 30 days x 55 lb DM/acre/day x 17 acres = 28050 lb DM

July: 0

Aug: 31 days x 45 lb DM/acre/day x 9 acres = 12555 lb DM

September and October: 0

4. <u>Consider the options throughout the period:</u>

- How much area will you offer the milkers each month?

April: graze all pastureland, plus hayland as needed.

May: 26 acres.

June: 33 acres.

July: graze all pastureland, plus hayland as needed.

August: 41 acres.

September and October: graze all pastureland, plus hayland as needed.

• Might you consider feeding conserved forage to the milkers? If so, how much and when?

April: a period of transition from winter feeding of conserved forage, so would continue feeding as required until all forage needs are met by pasture.

July: if pasture deficit develops due to dry conditions, feed as needed forage that was conserved in May and June.

September: instead of conserving surplus forage by machine harvesting in August, stockpile it in the pasture for grazing now. Forage quality and stand density won't deteriorate very much by stockpiling at this time.

October: begin transition of feeding conserved forage for winter.

• What feeding plan will you use for dry cows?

Milking cows graze to a residual of 1400 lb DM/acre; dry cows follow milkers through paddocks, grazing to a residual of 1000 lb DM/acre; will supplement dry cows with forage conserved during May and June as needed. (To be certain that there will be enough feed for dry cows, it is best to include estimates of their pasture energy and dry matter demand in the feed budget.)

• How much forage will you conserve and when?

May: conserve 52080 lb DM (26 ton DM)

June: conserve 28050 lb DM (14 ton DM)

August: stockpile 12555 lb DM (6.3 ton DM)

TABLES

Energy requirements are based on forage dry matter and are expressed as percent (Table 8-5 only) or pounds of total digestible nutrients (TDN), Mcal of metabolizable energy (ME)/lb DM, or Mcal of net energy lactation (NEL)/lb DM.

Energy values in the tables are the amounts of energy needed by animals and provided by pasture forage. So the values can be looked at in two ways:

1. For a given level of production, an animal of a given weight and physiological state needs a certain amount of daily energy, which can be obtained from certain pasture classes, based on their available energy contents.

2. A given pasture class, based on its available energy content, can support certain production levels of certain animal weights and physiological states.

TABLE 8-5. Pasture classification based on forage energy content, (DM basis) expressed as TDN, ME, and NEL.

Class	Energy values		
	TDN	ME	NEL
	%	----Mcal/lb----	
Poor	49.8	0.8	0.50
Fair	60.9	1.0	0.62
Good	71.9	1.2	0.75
Excellent	83.0	1.4	0.87

Note: energy values are range midpoints.

TABLE 8-6. Daily energy (NEL) needs of milking cows, according to liveweight (lb), milk and butterfat (%BF) production, and pasture class.

Live-weight & %BF	Pasture class	Milk production, lb/day							
		0	15	25	35	45	55	65	75
		----------------------------NEL-----------------------							
	Poor	7.5	16.8	----	----	----	----	----	----
800	Fair	7.3	15.2	21.4	----	----	----	----	----
4.9%	Good	7.0	14.1	19.5	25.4	32.3	----	----	----
	Exc.	6.8	13.9	18.8	24.1	29.8	36.1	----	----
	Poor	9.7	17.4	23.8	----	----	----	----	----
1100	Fair	9.5	16.1	21.2	26.7	----	----	----	----
3.8%	Good	8.9	15.2	19.4	23.8	28.8	30.2	----	----
	Exc.	8.8	12.8	17.0	21.0	25.4	29.8	37.0	----
	Poor	10.6	17.4	22.9	----	----	----	----	----
1300	Fair	10.1	16.3	20.7	25.6	31.3	----	----	----
3.6%	Good	8.4	15.4	19.6	23.6	28.9	34.1	39.8	----
	Exc.	8.3	15.0	19.4	22.7	27.8	32.4	35.7	41.6

TABLE 8-7. Daily energy (TDN) needs of growing beef cattle or dairy heifers, according to liveweight, pasture class, & weight gain.

Class	Weight gain, lb/day						
	0	0.55	1.10	1.65	2.20	2.75	3.30
	----------------------------TDN----------------------						
Liveweight = 450 lb							
Poor	6.5	7.9	9.7	11.8	----	----	----
Fair	6.2	7.3	8.7	10.2	12.3	----	----
Good	6.0	6.9	8.0	9.2	10.8	12.5	----
Excellent	5.8	6.5	7.5	8.5	9.8	11.2	13.0
Liveweight = 650 lb							
Poor	7.9	9.6	11.5	14.1	----	----	----
Fair	7.7	9.0	10.5	12.3	14.7	----	----
Good	7.3	8.4	9.6	11.0	12.9	15.3	----
Excellent	7.1	8.0	9.1	10.2	11.7	13.5	15.8
Liveweight = 850 lb							
Poor	9.3	11.3	13.5	16.6	----	----	----
Fair	9.0	10.6	12.3	14.5	17.2	----	----
Good	8.7	9.9	11.3	13.1	15.2	18.0	----
Excellent	8.3	9.5	10.6	12.0	13.8	16.0	18.8

TABLE 8-7. (continued). Daily energy (ME) needs of growing beef cattle or dairy heifers, according to liveweight, pasture class, & weight gain.

Class	Weight gain, lb/day						
	0	0.55	1.10	1.65	2.20	2.75	3.30
	--------------------------ME--------------------------						
Liveweight = 450 lb							
Poor	11.2	13.7	16.8	20.5	----	----	----
Fair	10.8	12.7	15.0	17.7	21.2	----	----
Good	10.4	12.0	13.8	16.0	18.6	21.7	----
Excellent	10.0	11.3	12.9	14.7	16.9	19.4	22.5
Liveweight = 650 lb							
Poor	13.7	16.6	19.9	24.4	----	----	----
Fair	13.3	15.5	18.1	21.2	25.4	----	----
Good	12.7	14.6	16.6	19.1	22.4	26.4	----
Excellent	12.2	13.8	15.7	17.7	20.3	23.4	27.3
Liveweight = 850 lb							
Poor	16.1	19.6	23.4	28.7	----	----	----
Fair	15.6	18.3	21.3	25.1	29.8	----	----
Good	15.0	17.2	19.5	22.6	26.3	31.1	----
Excellent	14.4	16.4	18.3	20.8	23.8	27.7	32.5

TABLE 8-8. Daily energy (ME) needed for ewes and lambs during lactation when grazing good quality pasture

Ewe liveweight	Period of lactation		
	Early	Mid	Late
90	6.2	5.5	4.8
120	7.9	6.9	5.5
150	8.6	7.6	6.2
Lamb pasture need	0.5	1.0	1.9

TABLE 8-9. Daily energy (TDN or ME) needed for maintenance of ewes of different liveweights (lb), according to pasture class.

Live-weight	Poor TDN	Poor ME	Fair TDN	Fair ME	Good TDN	Good ME	Excellent TDN	Excellent ME
90	1.2	2.1	1.2	2.0	1.1	1.9	1.0	1.8
120	1.4	2.4	1.3	2.3	1.3	2.2	1.2	2.1
150	1.5	2.6	1.4	2.5	1.4	2.4	1.3	2.3

TABLE 8-10. Daily energy (TDN or ME) needs of weaned lambs, according to liveweight (lb), pasture class, and weight gain.

Live wt.	Pasture class	0 TDN	0 ME	0.11 TDN	0.11 ME	0.22 TDN	0.22 ME	0.44 TDN	0.44 ME	0.66 TDN	0.66 ME
	Poor	0.8	1.3	1.2	2.1	---	---	---	---	---	---
45	Fair	0.8	1.3	1.0	1.8	1.4	2.4	---	---	---	---
	Good	0.7	1.2	1.0	1.7	1.3	2.2	2.0	3.5	---	---
	Exc.	0.6	1.1	0.9	1.6	1.2	2.0	1.7	3.0	2.5	4.4
	Poor	1.0	1.7	1.4	2.5	---	---	---	---	---	---
65	Fair	1.0	1.7	1.3	2.3	1.7	3.0	---	---	---	---
	Good	0.9	1.6	1.2	2.1	1.6	2.7	2.4	4.1	---	---
	Exc.	0.9	1.5	1.2	2.0	1.4	2.5	2.1	3.6	2.9	5.0
	Poor	1.2	2.1	1.7	3.0	---	---	---	---	---	---
85	Fair	1.2	2.0	1.6	2.8	2.0	3.5	---	---	---	---
	Good	1.1	1.9	1.4	2.5	1.8	3.1	2.7	4.6	---	---
	Exc.	1.0	1.8	1.3	2.3	1.7	2.9	2.3	4.0	3.2	5.5

TABLE 8-11. Daily energy (TDN or ME) needs of goats, according to liveweight (lb), weight gain (lb/day), and physiological state.

Physiological state and liveweight	Weight gain	Energy need TDN	ME
Wethers and dry does			
50	0-0.15	1.4	2.4
60	0-0.10	1.4	2.5
80	0-0.05	1.6	2.8
100	0	1.7	3.0
120	0	1.7	3.0
Pregnant does (last 8 weeks)			
50	0.38	2.0	3.4
60	0.35	2.0	3.5
80	0.33	2.2	3.8
100	0.30	2.4	4.1
Nursing does			
50	-0.05-0	2.1	3.6
60	-0.05-0	2.2	3.8
80	-0.05-0	2.4	4.1
100	-0.05-0	2.5	4.3
Growing kids and yearlings			
20	0.30	1.2	2.0
40	0.25	1.4	2.5
60	0.20	1.7	3.0
80	0.10	1.8	3.1
Developing bucks			
80	0.30	2.3	4.0
100	0.20	2.4	4.1
120	0.10	2.3	3.9

TABLE 8-12. Daily energy (TDN or ME) needs of milking goats per 2 pounds (1 quart) of milk produced, according to milk butterfat content (BF%).

Milk BF%	Energy need TDN	ME
3.0	0.69	1.19
3.5	0.73	1.27
4.0	0.79	1.36
4.5	0.84	1.45
5.0	0.89	1.54
5.5	0.94	1.63
6.0	0.99	1.72

TABLE 8-13. Daily energy (TDN or ME) needs of mature horses, pregnant mares, and lactating mares, according to degree of work and liveweight (lb).

Degree of work or physiological state	Energy needed at liveweights of 450		900		1100		1300	
	TDN	ME	TDN	ME	TDN	ME	TDN	ME
Horse at rest (maintenance)	3.9	6.7	6.6	11.4	7.7	13.4	8.9	15.4
Horse doing light work (2 hrs/day)	7.3	12.7	8.7	15.1	10.4	18.0	12.0	20.8
Horse doing medium work (4 hrs/day)	6.2	10.8	11.3	19.5	13.6	23.5	16.0	27.6
Mare, last 90 days of gestation	4.1	7.1	7.1	12.2	8.3	14.3	9.5	16.4
Mare, peak lactation	7.2	12.5	11.6	20.0	13.1	22.6	14.2	24.6

TABLE 8-14. Daily energy (TDN or ME) needs of growing horses, according to final mature weight (lb), age (mo), liveweight (lb), and weight gain (lb/day).

Mature weight & age	Live- weight	Weight gain	Energy need TDN	ME
450-lb mature weight				
3	100	1.54	3.5	6.1
6	200	1.10	4.0	7.0
12	300	0.44	3.8	6.6
18	350	0.22	3.8	6.6
42	450	0.00	3.9	6.7
900-lb mature weight				
3	200	2.20	4.9	8.5
6	375	1.43	5.6	10.2
12	575	0.88	6.5	11.2
18	725	0.55	6.7	11.6
42	900	0.00	6.6	11.4
1000-lb mature weight				
3	250	2.43	5.7	9.9
6	500	1.76	7.3	12.6
12	700	1.21	8.0	13.8
18	900	0.77	8.1	14.1
42	1000	0.00	7.7	13.4
1300-lb mature weight				
3	300	2.76	6.7	11.6
6	600	1.87	8.1	14.1
12	850	1.32	9.0	15.5
18	1000	0.77	9.1	15.7
42	1300	0.00	8.9	15.4

TABLE 8-15. Daily energy (TDN or ME) needs of growing and breeding pigs, according to liveweight (lb) and physiological state.

Liveweight & physiological state	Energy need TDN	ME	Liveweight & physiological state	Energy need TDN	ME
Growing pigs			Lactating gilts		
10-20	1.2	2.0	300-450	9.1	15.8
20-40	2.4	4.2			
45-80	3.1	5.4	Lactating sows		
80-130	4.6	7.9	450-550	10.0	17.4
130-220	6.4	11.1			
			Young boars		
Bred gilts			250-400	4.6	7.9
240-350	3.6	6.3			
			Adult boars		
Bred sows			400-550	3.6	6.3
350-550	3.6	6.3			

TABLE 8-16. Daily energy (TDN or ME) needs of chickens, according to physiological state and age.

Physiological state & age	Energy need TDN	ME
Broilers		
0-3 weeks	0.82	1.42
3-6 weeks	0.84	1.46
6-9 weeks	0.86	1.48
Replacement pullet layers		
0-6 weeks	0.77	1.34
6-12 weeks	0.79	1.36
12-18 weeks	0.79	1.36
18 weeds-laying	0.79	1.37
Layers	0.75	1.30

TABLE 8-17. Daily energy (TDN or ME) needs of growing turkeys, according to sex and age (weeks).

Sex & age	Energy need	
	TDN	ME
Male		
0-4	0.73	1.27
4-8	0.76	1.32
8-12	0.79	1.36
12-16	0.81	1.41
16-20	0.84	1.45
20-24	0.87	1.50
Female		
0-4	0.73	1.27
4-8	0.76	1.32
8-11	0.79	1.36
11-14	0.81	1.41
14-17	0.84	1.45
17-20	0.87	1.50

TABLE 8-18. Daily energy (TDN or ME) needs of ducks and geese, according to physiological state.

Physiological state	Energy need	
	TDN	ME
Ducks		
Starting	0.76	1.32
Growing	0.76	1.32
Breeding	0.76	1.32
Geese		
Starting	0.76	1.32
Growing	0.76	1.32
Breeding	0.76	1.32

REFERENCES

Chase, L.E., and C.J. Sniffen. 1985. Equations used in ANALFEED, a Visicalc template. Cornell Dept. Animal Science. Ithaca, New York. Mimeo.

Ensminger, M.E., and C.G. Olentine, Jr. 1978. *Feeds and Nutrition -- Complete.* Ensminger Pub. Co., Clovis, California. 1417 p.

Flanagan, J.P., J.P. Hanrahan, and P. O'Malley. 1987. *Lowland Sheep Production - Blindwell System.* An Foras Taluntais. Western Research Centre, Belclare, Tuam, Ireland. Sheep Series No. 2. 20 P.

Holmes, C.W. 1987. Pastures for dairy cows. p. 133-143. In. A.M. Nicol (ed.) *Feeding Livestock on Pasture.* New Zealand Society of Animal Production. Hamilton. Occasional Publication No. 10.

Holmes, C.W., and G.F. Wilson. 1984. *Milk Production From Pasture.* Buttersworth of New Zealand. Wellington. 319 p.

Mayne, C.S. 1991. Effects of supplementation on the performance of both growing and lactating cattle at pasture. British Grassland Society Occasional Symposium No. 25, p. 55-71.

Mayne, C.S. (according to A. Nation) 1992. Supplemental feeding found to be of little value. *The Stockman Grass Farmer.* 49(7):1,7.

Milligan, K.E., I.M. Brookes, and K.F. Thompson. 1987. Feed planning on pasture. p. 75-88. In. A.M. Nicol (ed.) *Feeding Livestock on Pasture.* New Zealand Society of Animal Production. Hamilton. Occasional Publication No. 10.

Nation, A. 1992. *Pasture Profits with Stocker Cattle.* Stockman Grass Farmer, P.O. Box 9607, Jackson, MS. 39286-9909. Phone: 800-748-9808. 192 p.

Rayburn, E. 1990. *Forage Quality of Intensive Rotationally Grazed Pastures - 1989.* Seneca Trail Resource Conservation and Development Area and Cornell Extension Service. Franklin, New York. 34 p.

9
Extending The Grazing Season

On the dead branch
A crow settles --
Autumn evening
Basho

The grazing season can be extended at either end or during the season, when stored feed would otherwise have to be fed because of a deficit of pasture forage. The best way to extend your grazing season is to apply Voisin management intensive grazing. This alone will add months of additional grazing time at usual season ends, and will minimize or eliminate plant production slumps that otherwise occur during the season. Special techniques to extend the season range from stockpiling surplus forage for winter or early spring grazing, to seeding forage brassicas for autumn and late fall grazing.

Using annually seeded crops to make up for pasture shortages incur costs of seedbed preparation, fertilizer, seed, fuel, machinery, labor, and time. So before seeding a crop, compare estimated costs with projected returns to determine if what you're thinking of doing will pay for your trouble. Ask Extension agents for suggestions about varieties and seeding rates and times suited to your conditions. Give animals small paddocks or narrow strips of these crops with portable fencing, so that most of the valuable forage is eaten. Annual crops for fall grazing should only be seeded on land that won't be at risk of soil erosion during winter after the cover has been grazed off.

AUTUMN

Autumn to early winter is the time that I think of first in trying to extend the grazing season. It's the time in Vermont when we can most easily extend the grazing season, because in early spring (March to mid-April) we have what is called Mud Season. There are many Vermont jokes about it, and schools still regularly schedule a Mud Season break. It goes back to the time before road pavement when people simply couldn't get to school from where they were in early spring. We still can barely get through our gravel road, let alone try to put livestock on pasture during Mud Season! Of course fall and summer also are cool and wet in Vermont, but it's the degree of wetness that matters.

Hay Aftermath
Besides improving grazing management, the next easiest and cheapest way of extending the grazing season is to graze autumn regrowth of hayland, rather than machine-harvesting and feeding it as green-chop or storing it. This is very simple to do with portable fencing.

At this time plant regrowth rate will be too slow to worry about regrazing before adequate recovery. But it's still best to provide just enough forage allowance for short grazing periods, by using small paddocks rather than having livestock wandering around large hay fields. By restricting animal movement, small paddock breaks save animal energy, reduce plant crown damage from treading (especially alfalfa), soil pugging or poaching (punching holes), and soil compaction. The higher stocking density within paddocks, compared to large fields, results in quick, more complete grazing of available forage. The combination of these effects from small paddock grazing results in higher production per acre, compared to the grazing of large areas. There shouldn't be any rejected forage, because there was no previous dung dropped.

Stockpiled Pasture
Surplus forage can be deferred from grazing (stockpiled) in autumn on part of the pasture area, and carried forward for grazing in fall, winter, or early spring. Irish researchers

calculated that every 3/4 inch of autumn-saved pasture forage spread over 50 acres and rationed out to grazing livestock, is equivalent to feeding out 50 tons of haylage! It works well to use about 1/3 of this accumulated forage during fall, 1/3 during winter to feed dry animals, and the other 1/3 for early spring milk production.

Stockpiling pasture forage may result in earlier spring growth by insulating plants and soil during winter. As a result, plants may not need to use as much of their food reserves, and the soil may not freeze as solid. The forage should be slowly grazed off to a postgrazing mass of about 1000 lb DM/acre by careful rationing. In areas where the insulating forage cover is removed as spring weather approaches, the soil warms quickly. Microorganisms become active early and begin to make nitrogen available, which promotes early spring growth of plants that came through the winter in good condition.

Stockpiling pasture and grazing it off during fall, winter, and spring staggers the amounts of forage present in paddocks. As a result the rapid growth of spring will be much easier to manage.

In using this method try to stockpile different parts of your pasture each year, so that the sward doesn't thin out from less grass tillering or loss of low-growing plants due to shading under high pasture masses. Orchardgrass, timothy, and bromegrass appear to tolerate stockpiling better than Kentucky bluegrass. Perennial ryegrass and Matua prairie grass grow well during September and October, and during that time can accumulate large amounts (about 2 tons DM/acre) of forage. With Matua, however, this results in winter killing and slow recovery in spring. Fall growth of Matua needs to be grazed down to a 3- to 5-inch residue every month in the fall, rather than allowed to accumulate.

Stockpiling pasture forage works especially well with tall fescue. If allowed to grow ungrazed from August until November, it can support beef cattle grazing during most or all of the winter. Freezing temperatures of fall cause changes in fescue plants that make them more palatable.

Crop Residues

Farmers traditionally graze corn fields after harvest with beef cattle to eat any of the crop that was missed. There's no reason why beef and other livestock can't graze all crop residues, including those of small grains. When properly supplemented, crop residues can form a valuable and inexpensive part of the ration. Plant parts left in the field can be almost as important nutritionally as those removed with a combine. For example, small grain (barley, oats, rye, triticale, wheat) straw contains about 5 percent protein and 37 percent energy. Grazing residues in the field helps decrease weed populations, and returns much of the crop organic matter to the soil. It certainly is a better solution than burning to remove unwanted residues. Check labels of any pesticides that were used on the crop to be certain that the residue can be grazed.

Small Grains

Small grains such as barley, oats, rye, triticale, and wheat can be seeded in early autumn to provide good grazing 6 to 8 weeks later or for grazing in late winter and early spring. Because small grains grow best under cool temperatures, they can be no-till drilled into short pasture sod in autumn after the last grazing of paddocks to provide early spring grazing. At this time pasture plant growth slows so much that there is little competition with the small grain seedlings. No herbicides are needed.

For example, a farmer in Virginia drills about 55 lb/acre of wheat or rye with 15 lb/acre of Austrian winter pea, or 50 lb/acre of oats with 15 lb/acre of Green Globe turnip seed into pasture sod in early fall for late winter and early spring grazing.

In Missouri a combination of autumn-seeded rye and hairy vetch is used to obtain excellent grazing until

February. After a 60-day recovery period, the rye and vetch mixture can be grazed again in April and May.

Successful use of small grains drilled into and grazed on pasture sod in the fall, winter, and spring depends on having a well-drained soil that remains frozen in winter. If the soil is too wet in fall and thaws during winter and early spring, the pasture may become badly damaged and consequently will not be very productive the following grazing season. Whenever soils are wet and soft, use techniques discussed under Wet Soil Conditions in chapter 5.

Brassicas

Brassicas such as turnips, rape, kale, and swedes can be very useful in cutting feed and feeding costs during mid- to late season. These cold-hardy, fast-growing species establish easily, thrive under many conditions, crowd out weeds, and can be grazed well into winter in many areas. Seed costs are low ($3 to $8/acre), and only moderate soil fertility is needed. Dry matter production is high (4 ton DM/acre for turnips and rape; 6 ton DM/acre for kale and swedes). They can be seeded anytime from late spring to midsummer, and be ready to graze in 6 weeks from seeding. Rape can be grazed in autumn and again later in the fall or early winter.

Forage Chicory

Forage chicory (e.g. Puna) provides a rich source of minerals, vitamins, and plant hormones needed by livestock. Its tap root penetrates soil up to 10 feet, recycling minerals and helping to break up hard layers and improve drainage. Chicory is a perennial plant, so it can provide grazing throughout the season.

Chicory may be frost-seeded or sown in early spring, late summer, or fall. It's best to seed in a prepared seedbed, but it establishes well by broadcasting on pasture sod in early spring just ahead of grazing livestock, using hoof action to cover the seed. Seed 2 lb/acre in a mixed grass-legume sward, or 5 lb/acre for a pure stand. Cover seed no deeper than 1/4 inch. Chicory will grow at soil pH ranging from 4 to 8, but does best at 5.5 to 6.

Annual Ryegrass
Tetraploid annual ryegrass (e.g. Concord Italian ryegrass) can help increase production on land that is being converted from tilled cropping to pasture. It appears to tolerate the poor soil growing conditions (e.g. herbicide residue, low soil organic matter and fertility, damaged soil structure, low soil organism populations) that generally exist following repeated corn cropping. On such land, it may be beneficial to grow only annual ryegrass for a few years until soil conditions improve, before drilling in perennial grasses and legumes. Depending on your local climatic conditions, Concord "annual" ryegrass may persist and thrive for 2 to 3 years. Seed 20 lb/acre early in spring or in late summer to get good grazing 4 to 6 weeks later. Begin grazing the ryegrass when it's about 6 inches tall. Carefully ration out the high-quality forage and closely back fence to keep animals from grazing regrowth. Depending on the weather at your location, the ryegrass will regrow for successive grazings.

Annual ryegrass also can be seeded in spring with an oat or barley crop for fall grazing. Just seed 15 to 20 lb/acre of ryegrass along with oats or barley seeded at the normal seeding rate of about 3 bushels/acre. The ryegrass seed can be mixed with the grain in the seeder. Harvest the oats or barley for grain and straw in July or August. Then begin grazing the ryegrass in September when it's about 6 inches tall. Ration out the forage and keep animals from grazing regrowth, so the ryegrass regrows for successive grazings.

MIDSUMMER

Warm-Season Perennial Grasses

Warm-season native grasses such as switchgrass, big bluestem, and Indiangrass can complement cool-season species such as Kentucky bluegrass. Cool-season grasses produce most of their forage by early to mid-June, just when warm-season grasses are reaching their period of greatest productivity.

To manage them correctly, warm-season grasses must be grown separately from cool-season species. Iowa and Missouri farmers have found that it works well to keep 25 to 40 percent of their pasture area in warm-season grasses, and the rest in cool-season species.

Warm-season grasses provide about a month of grazing in midseason, and another grazing after they have regrown. The amount of grazing time from warm-season grasses can be extended by seeding species separately that reach top production levels at different times, probably because of varying temperature and light requirements. For example, in Missouri switchgrass begins producing a lot of forage in late May to early June, big bluestem reaches peak production in July, followed by Indiangrass in August to early September.

Warm-season grasses should be grazed when they are 15 to 18 inches tall, and only grazed down to 6 to 8 inches tall. They don't recover well if grazed off close to the ground, and will persist only if allowed to recover adequately after their midsummer productive period.

Manure or nitrogen fertilizer should be applied to warm-season grass swards, because they are so dense and tall that they choke out most legumes that otherwise would make nitrogen available to the grasses.

Warm-Season Annual Grasses

If your pasture production usually declines so much in midsummer that you need supplemental feed, you can seed annual grasses such as corn, sudangrass, forage sorghum, and pearlmillet in advance to provide grazing during that time. These grasses grow rapidly, can withstand considerable moisture stress, and produce high total forage yields of 3 to 6 ton DM/acre.

Corn

Direct grazing of standing corn is a common practice in areas of New Zealand that are dry during summer. Corn has proven to be a reliable alternative to parched cool-season grasses.

Drill seed in 7-inch rows in a firm seedbed at a seeding rate of 40-50,000 kernels/acre. Corn for grazing also can be broadcast with a spin seeder on a prepared seedbed at 50,000 kernels/acre, followed by a cultipacker to cover the seed. These rates result in the most bulk of leaf, with less thick stalks. Sow seed at least twice, 4 to 6 weeks apart to get a succession of green, leafy plant material over 60 to 65 days of summer grazing. May 1, June 1, and July 1 work well for planting dates in the Midwest.

Use ammonium nitrate or sulfate to apply 150 lb N/acre at seeding. Don't use urea because it can lose about half of its nitrogen and kills earthworms.

Plan to finish grazing of the plants before they develop full flower. Careful timing of seeding and beginning grazing at first sign of tasseling results in a plentiful, nutritious green feed during what otherwise would be a difficult feeding period.

Open-pollinated corn varieties work best because of their slower maturity and better drought tolerance, but hybrid silage-type varieties also perform well. Consider using Baldridge grazing maize that was developed specifically for grazing. It yields about 15 ton DM/acre, containing 10-13% crude protein, 67% TDN, and 0.68 Mcal NEL/lb DM. Baldridge is available from Old Field Seed Co., Winchester, KY; phone: 800-448-5145.

Carefully ration out the forage, using long narrow

breaks about 3 feet wide. Use polytape to form breaks so animals can see the fence. Animals will be shocked if the tape touches any corn plants, so you'll have to knock down or cut off the plants where the tape is placed.

Make the transition gradually to grazing corn over 10 to 14 days to allow rumens time to adjust. Milk production may increase when cows graze standing corn, if they're also grazing grass and clover mixtures that contain a high level of protein, or receive a protein supplement. Corn doesn't contain enough protein if used alone for lactating cows. Cows should graze corn only for 3 or 4 hours in the morning, followed by grazing other pasture during the rest of the day; this provides needed protein and prevents the problem of the animals not seeing the fence at night in the corn.

Standing corn can be used alone for grown beef stockers that are being fattened for slaughter. Stockers weighing less than 850 pounds need to alternately graze grass-legume swards to get enough protein.

Sudangrass, Forage Sorghum, And Pearlmillet
The highest forage yields are gotten from these grasses if they're allowed to grow to 30 to 48 inches tall before grazing. At that height only very narrow strips of the forage can be offered to livestock to minimize waste. Most uniform paddock grazing with least waste is achieved by grazing these grasses when they're 8 to 12 inches tall. Regrowth of summer annual grasses depends greatly on amount of leaf surface, and presence and growth of meristematic buds left on the stubble. So don't graze these grasses lower than 4 to 6 inches from the soil surface.

Be careful when grazing sudangrass and sorghum-sudangrass hybrids so that your livestock don't become poisoned by hydrocyanic acid that may form in these plants. Hydrocyanic acid potential differs with variety, stage of growth, level of soil nitrogen, and environmental conditions. Hybrids tend to form hydrocyanic acid more than sudangrass. Higher concentrations of hydrocyanic acid develop in new growth of grass, at high levels of nitrogen fertilization, and under stress conditions of drought or frost. So apply only low amounts of manure or nitrogen fertilizer for these grasses, and don't graze them at early growth stages or right after a drought or frost.

EARLY SPRING

As discussed above, grazing of pasture or autumn-seeded small grains can be deferred or stockpiled in the fall so that the forage can be carried over winter for early spring grazing. Stockpiled forage maintains nutritional quality, especially if it's overwintered under snow and freezing temperatures. Winter wheat, rye, and triticale can regrow after fall grazing, and provide forage for early spring grazing.

REFERENCES

Brown, C.S., and J.E. Baylor. 1973. Hay and pasture seedings for the Northeast. p. 437-447. In M.E. Heath, D.S. Metcalfe, and R.F. Barnes (eds.) Forages. Iowa State University Press. Ames, Iowa.

Cramer, C. 1989. 12 ways to make pastures really produce. The New Farm. May/June. p. 25-29.

Cramer, C. 1989. "Offbeat" feeds save 20-35%. The New Farm. September/October. p. 30-33.

Cramer, C. 1990. "Grass farming" beats corn! The New Farm. September/October. p. 10-16.

Fribourg, H.A. 1973. Summer annual grasses and cereals for forage. p. 344-357. In M.E. Heath, D.S. Metcalfe, and R.F. Barnes (eds.) Forages. Iowa State University Press. Ames, Iowa.

Henning, A.D. 1992. Autumn saved pasture: the starting point for next season. *Stockman Grass Farmer*. 49(9):14.

Johnson, R. 1996. Forage chicory: a drought-tolerant addition to pasture salad. *Stockman Grass Farmer*, 53(8):17-18.

Jung, G.A. 1993. A summary of new research on extending the grazing season. *Stockman Grass Farmer.* 50(9):8-10.

Jung, G.A., R.L. Reid, L.C. Vona, and L.P. Stevens. 1987. Grazing systems, herbage quality, and animal behavior on warm-season and cool-season grass pastures on hill country in the northeastern United States. p. 45-63. In F.P. Horn, J. Hodgson, J.J. Mott, and R.W. Brougham (eds.) *Grazing Lands Research at the Plant-Animal Interface.* Winrock International. Morrilton, Arkansas.

Knight, A. 1988. Making milk for $2.50. American Agriculturalist. November. p. 15-16.

McIver, D. 1991. Quadruple your stocking rate. The New Farm. May/June. p. 17-21.

Nation, A. 1992. Fall-seeded rye pasture can cut hay bill. *Stockman Grass Farmer.* 49(10):6.

Nation, A. 1992. Importance of stockpiled pasture; Al's Observations. *Stockman Grass Farmer.* 49(10):21.

Nation, A. 1993. Grazed corn can spell summer pasture. *Stockman Grass Farmer.* 50(9):10.

Nation, A. 1993. Fall grass can reduce need for hay. *Stockman Grass Farmer.* 50(9):18-19.

Nation, A. 1993. The more winter feeding options, the better. *Stockman Grass Farmer.* 50(10):1,6-7.

Nation, A. 1994. New high protein grazing corn available. *Stockman Grass Farmer.* 51(2):37.

Nation, A. 1995. Grass rules apply to grazed corn. *Stockman Grass Farmer.* 52(9):1,4-6.

Nation, A. 1996. Wisconsin graziers discover benefits of spring-planted annual ryegrass. *Stockman Grass Farmer.* 53(12):1.

O'Sullivan, M. 1982. Set stocking or rotational grazing? Farm & Food Research. An Foras Talúntais. Johnstown Castle Research Center, Wexford, Ireland. April. p. 43-45.

Reid, R.L., and G.A. Jung. 1973. Forage-animal stresses. p. 639-653. In M.E. Heath, D.S. Metcalfe, and R.F. Barnes (eds.) Forages. Iowa State University Press. Ames, Iowa.

Zahradnik, F. 1986. Bridge the midsummer pasture gap. The New Farm. January. p. 10-13.

10
Fine Tuning

The flowers depart
When we hate to lose them;
The weeds arrive
While we hate to watch them grow.

Dogen

I left this management topic until last; that's where it should be. After you've done everything to increase plant and animal yield by improving grazing and livestock management, and are using the forage being produced as much as possible, then consider additional inputs such as renovation, nitrogen fertilization, and irrigation -- if you want to further increase productivity.

PASTURE IMPROVEMENT

Conventional Renovation

Whenever anyone mentions pasture improvement in the United States, almost everyone thinks of renovation. This process as usually understood includes at least partially destroying the sod, plus fertilizing, liming, seeding into a prepared seedbed or killed sod, and applying herbicide to control weeds in the new seeding. In short, doing whatever is needed to establish desirable plant species. Destroying the sod involves either 1) severe overgrazing to weaken the sod, followed by several passes with a heavy disc at different times of the year to cut the sod into pieces, or 2) applying herbicides that kill all or most of the plants, including desirable species.

In the instructions for this kind of renovation, some suggestion may be made to change the grazing management so the pasture doesn't revert to the same mess that it was before. But the suggestions generally have been vague and involved using a few large paddocks, long occupation periods, and fixed recovery periods. This kind of management doesn't take into account the needs of plants, and consequently fails to maintain renovated pastures in an improved condition.

Is renovation as conventionally attempted, practical or even necessary?

Conventional renovation practices essentially are attempts at transferring field-cropping techniques to a pasture situation. They ignore the fact that a pasture is a different ecological environment, compared to that of monoculture field crops grown on plowed and cultivated soil. Destroying a pasture sod and plowing and cultivating soils, drastically changes conditions in soils and disrupts balanced relationships among the organisms that live there and make soils alive. It takes a long time after plowing a pasture sod for the relationships among soil organisms to reach the equilibrium, and for the soils to return to the same conditions, that existed before plowing. Andre Voisin wrote that it could take about 100 years -- if the pasture is well managed after renovation!

Andre Voisin may have been the first person to question the practice of pasture renovation that destroys the existing sod, when he wrote that plowing a pasture does not make up for defective grazing management. Voisin stated that the logical approach to improving pastures is to change grazing management and wait a few years, to see how the pasture plant community responds. He cited observations by other ecologists and results of research studies, which clearly showed that pasture plant communities are extremely dynamic, changing very rapidly with changes in grazing management.

One study cited by Voisin is especially interesting because it was done at Cornell University, and should have influenced thinking about pasture improvement in the United States. The Cornell study involved pasture plots all sown with the same simple mixture of Kentucky bluegrass and white clover. Plots were cut to 1/2 inch above the soil surface at different fixed intervals. After two years of cutting treatments, clover contents in the plots ranged from 1 to 80 percent, depending on the cutting management used. This was one study among many others done elsewhere, which showed that the botanical composition of pasture swards depends mainly on recovery periods between cutting or grazing.

There's good evidence that existing pasture plants can be very productive if managed well. For example, recent research in Scotland and England showed that unless a farmer is willing to apply more than 110 lb N/acre/year, renovation to introduce new varieties of perennial ryegrass can't be justified. If managed well, grasses (e.g. bromegrass, timothy, orchardgrass, Kentucky bluegrass, and quackgrass) already existing in permanent pastures produce just as much forage and protein per acre as new selections of perennial ryegrass without the expense of reseeding and heavy nitrogen fertilization.

Besides, I have yet to see anyone in the United States fully using the increased amount of forage that's produced when Voisin management intensive grazing is applied. There's no sense in spending money, time, and labor to renovate a pasture to promote higher production, if the forage that's already being produced isn't eaten.

Another aspect ignored by advocates of conventional renovation is that most permanent pastureland in the northern United States is so-called marginal land. The reason it's considered to be marginal (and is in permanent pasture) is because it has steep, rocky, or shallow soils, ledge outcroppings, boulders, brush, or trees that limit or prevent its use for cultivated field crops. These problems make the field-cropping techniques of conventional renovation very difficult or impossible to use on most permanent pastureland that needs improving.

Rational Renovation
These methods involve thoughtful ways of improving pasture swards without disrupting the plant, soil, and soil

organism balanced relationships needed for sustained pasture production.

Overseeding (Frost or Sod Seeding)

Frost seeding takes advantage of the freezing and thawing action that opens and closes soil pores in the early spring. The soil movement covers seed that simply was broadcast on the frozen soil just before or right after the snow melts.

My earliest attempt at improving pasture swards involved frost seeding white clover in an old meadow, without seedbed preparation. The meadow was then grazed with sheep under intensive management with multiple paddocks. By October, clover content in the plant community had increased from its original 2.5 percent, to about 25 percent where clover had been seeded.

But to get to the frost-seeding experiment, I had to walk across pasture areas that had been grazed under intensive management for 3 years with dairy heifers. I noticed that in those areas white clover content had increased to about 30 percent, without any kind of seeding or any other treatment except changing the grazing management! By the end of the second year of the frost-seeding experiment, white clover content in all plots had increased to about 30 percent, regardless if clover had been seeded or not.

This experience with pasture sward improvement was the first I had during several years of observing the desirable changes in plant composition and increases in forage yield that generally occur in Vermont pastures when management intensive grazing is used. These observations persuaded me that the conventional method of attempting to improve pastures (i.e. renovate and then maybe perhaps change the grazing management) is just the opposite of what it should be.

There are, of course, situations where little legume or desirable grass seed might exist in the soil, such as in long abandoned pasture, old tall-grass hayfields, or tilled cropland where herbicides have been applied. On tilled cropland you can seed whatever grasses and legumes you want into a prepared seedbed. Where a sod already exists, you can speed the transition to a more productive sward by overseeding the legumes, grasses, and forbs (e.g. Puna chicory) that you want to introduce. It especially may be beneficial to overseed different areas each spring to introduce new legume varieties, because they probably can fix more nitrogen than those existing in your pasture. If swards are thinned by drought and winter killing, overseeding of grasses and legumes should be done every year. Annual overseeding is a small price to pay, given the value of high-quality pasture forage.

Frost seeding won't work if the soil surface is covered with a mat of undecomposed plant residue that prevents good soil and seed contact. In this case, you can use a sod-seeder or change your grazing management for at least one season until the residue mat disappears, before attempting to seed anything.

Frost-seeding success also varies with soil type, plant species, and competition from existing pasture plants. Heavy, moist soils heave more and may cover seed better than coarse textured, drier soils. Legumes are easier to frost seed than grasses. It's difficult to introduce new grass varieties into an existing, vigorous pasture sward. For example, I've introduced red and white clover into my existing swards easily with frost seeding, with or without passing animals (sheep) over frost-seeded areas for hoof action to press in the seed. But I haven't been able to establish grasses (perennial ryegrass, Matua prairie grass, tall fescue, Kentucky bluegrass) very well using the same method. It probably indicates that existing grasses are well adapted to growing conditions and management, and are very competitive against new seedlings.

Mob Stocking
Pastureland that is overgrown with brush and weeds can

be improved quickly and inexpensively by applying the technique of mob-stocking. Fence in small areas of the rough land and confine large numbers of nonproducing animals (e.g. dry cows, goats, or ewes) to each small area until they have eaten all of the leaves that they can reach. The areas must be small enough so that they can be grazed completely in 24 hours or less. Every 2 to 3 days return the animals to better pasture for 1 to 2 days so that they don't become stressed for nutrients. Water, salt, and minerals must be available at all times to the animals.

Begin grazing early in spring just after leaves form on the brush, and graze repeatedly every time the leaves regrow during the season. Before beginning to graze, carefully check the area to make certain that no poisonous plants are present that could harm your livestock when grazed. If there are poisonous plants, remove them from the area. Mob-stocking can change rough land into productive pasture within a few years. This change will be even quicker if you overseed grasses and legumes in spring, and mow the area with a rotary brush-cutting mower, either in early spring or in midsummer when weedy plants are forming flower buds.

Nitrogen Fertilizer

Nitrogen fertilizer can be used to extend grazing at the beginning and end of the season. Knowing when and how much to apply so that fertilizer isn't wasted is part of the art of pasture management. Voisin used stages of growth of nearby trees to indicate when to apply nitrogen to his pasture, and close observation to determine the amounts of nitrogen to apply. You can do the same for your farm.

In the early spring, Voisin applied nitrogen to his pasture when wild cherry trees had barely started to bud. He applied 33 lb N/acre to the first two paddocks that he planned to graze, 27 lb to the next two paddocks, 20 lb to the next two, and 13 lb to the next two. He didn't apply nitrogen to the rest of the pasture. This timing and application amounts provided earlier grazing and staggered plant growth, making the spring flush of growth easier to manage.

To extend grazing into the fall, Voisin suggested applying 14 lb N/acre in late July-August, 20 lb in August-September, 27 lb in September-October, and 27 lb in October-November. Timing and amounts of application need to be adjusted to local conditions.

By today's standards, Voisin was remarkably conservative in the amounts of nitrogen that he applied. He certainly followed the adage that a little bit often is better than a lot infrequently. He marked 16- x 16-foot unfertilized areas with stakes and string within his pasture so he could see and compare responses to fertilizer applications. The amounts of nitrogen that he applied apparently were enough for his well-developed pasture to provide the increased forage yield that he wanted.

Today almost no one would consider anything less than 50 lb N/acre/application. In some countries in Europe farmers apply as much as 400 lb N/acre/year to pasture! This practice is ending, however, because excess nitrogen is polluting surface and ground waters, besides wasting nonrenewable resources. I suggest that you try to use similar timing and amounts that Voisin used, and see what happens.

Whatever you do, be careful! Excessive, wasteful nitrogen fertilization unnecessarily uses nonrenewable resources, pollutes surface and ground waters, substitutes expensive fertilizer nitrogen for free legume-fixed biological nitrogen, and wastes your money. Nitrogen-fertilized forage is more lush and may cause bloating. Don't apply nitrogen fertilizer if you aren't prepared to use the extra forage that will be produced! Nitrogen-fertilized areas must be grazed closely and probably more frequently to prevent grasses from crowding out white clover. For fertilizer, use ammonium nitrate or ammonium sulfate, not urea, to avoid killing earthworms and losing volatilized nitrogen.

Irrigation

If your area tends to have dry spells during the growing season, it might be worthwhile to consider irrigating your pastureland. Pasture irrigation actually can be more cost-effective in relatively high rainfall areas than in arid areas. This is because irrigation provides the most benefit when used to supplement rainfall, rather than as the only source of water.

Precision irrigation developed in Tasmania by Gerard Van Den Bosch aims at keeping pasture soil constantly between saturation and wilting points, to achieve high plant growth rates all season. Wet and dry soil conditions are avoided and so are problems of soil compaction, pugging, and salt buildup. In this method, timers are used to apply a small amount of water frequently and always at night to reduce evaporation loss and wind interference, and to use off-peak electrical rates for pumping. The Van Den Bosch system uses a series of light-weight sprinklers on skids, attached to flexible, kink-resistant, lateral-line hoses that attach to hydrants on main lines. The system applies 1 to 1.5 inches of water per 8 hours from streams or reservoirs built to accumulate rainwater for irrigation. For more information contact Bosch Irrigation Ltd., PO Box 1420, Whangarei, NZ. Fax: 64-9-438-5099.

New Seeding

It can take several years to develop a sod on land that was tillage cropped with herbicides, depending upon how many years the soil was tilled and the farming methods that were used. Usually soil structure has been damaged, soil organic matter has been depleted, and there is herbicide residue. It's not simple nor easy to change from bare soil in this condition to a well-developed, dense, productive pasture sward.

If you are planning to seed pasture plants on land that has been in tilled crops, stop using herbicide at least 2 years before attempting to seed it. If the soil needs liming, apply lime at least a year before seeding, to allow time to get rid of any herbicide that's released by raising soil pH.

Farmers in Pennsylvania have managed to seed pasture plants following silage corn by seeding winter rye after the corn is harvested. In late winter they spin broadcast grass and legume seed into the rye. In spring they graze the rye several times with cows at high stocking density; hoof action covers the grass and legume seeds.

Legume Seed Inoculation

When overseeding a sod or seeding a prepared seedbed, always inoculate legume seed. The rhizobia bacteria in inoculants probably are capable of fixing more nitrogen than rhizobia in the soil.

Despite directions usually found on inoculant packages, dry application of inoculant doesn't work. Only about 20% of the dry material sticks to seeds, and survival of rhizobia on seeds decreases when applied dry.

Apply inoculants with a sticker. Use a 25% sugar solution, 10% corn syrup solution, or a commercial preparation (e.g. Nitragin's Pelinoc). These materials increase survival of rhizobia on seeds. Other sticky substances, such as maple syrup or soda pop may contain compounds that kill rhizobia.

Here's one way to inoculate seed:

1. Place 25 pounds of seed in a clean tub.

2. Prepare sticker:

Sugar: dissolve 1/4 cup of sugar in a cup of water

Corn syrup: add 1 ounce of syrup to 9 ounces of water

3. Add 1/3 of an inoculant package to 7 ounces of the stickers to make a slurry.

4. Add the slurry to the seed and mix well.

5. Add the rest of the inoculant package to completely coat and dry the seeds. This is four times the rate recommended on the inoculant package, but is necessary for best results. It isn't possible to over-inoculate.

6. Air dry the seeds by spreading them out in the shade. Add more inoculant or finely ground limestone to speed up the drying.

7. Inoculate in the shade, never in direct sunlight, because ultraviolet rays in sunlight kill rhizobia.

8. Broadcast the seed as soon as possible after inoculating.

Soil Erosion

Tillage cropping on sloping land results in severe soil erosion that removes topsoil and leaves gullies. Under permanent pasture cover, topsoil can be rebuilt and gullies can be healed. Joel Salatin has developed a simple and effective way of healing gullies: 1) Place brush or trees, root ends facing downslope, in the lower end of the gully to catch and hold soil and plant debris. 2) Fill the upper end of the gully with stones to slow flowing water.

Conclusion

Voisin was right! Pasture renovation won't make up for defective grazing management. In general, to improve your pasture:

• Don't plow, cultivate, or kill the pasture sod.

• Test pasture soils and plant tissues, and correct major soil fertility and pH problems.

• Subdivide the pasture into as many paddocks as you think necessary to get the recovery periods needed for your area.

• Follow the Voisin management intensive grazing method as closely as possible, making changes that seem necessary to adapt it to your local conditions.

• Overseed by frost seeding or using a sod seeder to thicken swards and introduce new grass, legume, and forb varieties.

• Be patient. Allow the pasture environment time to recover from the many years of abuse that it has endured.

REFERENCES

Clark, E.A. 1990. The principles and tools of intensive pasture management systems. Proceedings Field Crops Expo 90, Acadia University, Wolfville, Nova Scotia. July 15-18.

Davies, A., W.A. Adams, and D. Wilman. 1989. Soil compaction in permanent pasture and its amelioration by slitting. Journal of Agricultural Science, Cambridge. 113:189-197.

Frame, J. 1988. Pasture management in Europe. Proceedings Pasture Management Workshop "Improved Production for Your Pasture", Nova Scotia Agricultural College, Truro, Nova Scotia.

Johnstone-Wallace, D.B. 1945. The principles of pasture management. Proceedings of the New York Farmers 1944-1945.

Jones, M. 1933. Grassland management and its influence on the sward. Journal of the Royal Agricultural Society of England. 94:21-41.

Kendall, D. 1988. Nature's no-till pastures. The New Farm. May/June. p. 16-18.

Murphy, W.M., D.T. Dugdale, D.S. Ross, and L.S. Dandurand. 1982. *Vermont Alfalfa Needs Inoculation.* University of Vermont Agricultural Experiment Station. Research Report 24.

Nation, A. 1994. Spring pasture management is the key to the whole year. *Stockman Grass Farmer.* 51(4):25-27.

Nation, A. 1994. Pasture species diversity stabilizes the forage curve. *Stockman Grass Farmer.* 51(9):1.

Nation, A. 1995. Establishing perennial pasture in cropland. *Stockman Grass Farmer*. 52(3):30-31.

Nation, A. 1995. Precision pasture irrigation is goal downunder. *Stockman Grass Farmer*. 52(6):1,4.

Traupman, M. 1990. Mechanical "earthworms". The New Farm. March/April. p. 12-13.

Voisin, A. 1959. *Grass Productivity*. Island Press, Washington, D.C.

11
Social, Environmental, & Economic Benefits Of Feeding Livestock On Well-Managed Pasture

I scooped up the moon
In my water bucket...
And spilled it on the grass

Ryuho

SOCIAL BENEFITS

Rural Culture

One out of four Americans who were farming 10 years ago have left the land. In 1900, more than 40% of our people lived in the country; less than 2% live there today. In that enormous migration to the cities, we lost a fundamental family structure that today only exists in the family farms that remain. It's what Wendell Berry calls the home economy, where family, home, and work are interrelated, not separated. In the home economy of the family farm, children and parents work together. The children learn from the parents in a process of education that goes back generations. No one else in America has the chance to learn from their parents on such a continuing and broad basis. This is something essential and good that is being lost with the migration of our people from the land.

Rural culture is absolutely essential to sustain civilization. Despite that, the people who are involved in rural culture are a tiny minority and their numbers are dwindling daily. They are struggling not only to maintain their farming livelihoods, but for acknowledgement of their very right and critical need of that livelihood by the overwhelming, largely apathetic and unsympathetic majority of the population (12).

Goal-less Problem Solving

Bedrock, desert, and ruins are all that remain of 26 civilizations that have been destroyed, mainly because their agricultural practices led to loss of biodiversity on which stability of land and water depends. First one species disappeared, then ten, then a hundred. The water cycle deteriorated. Then the soil blew or washed away. The rural cultures suffered and died. Then the villages disappeared, followed by the towns. Finally all that was left were the lonely cities, and eventually they disappeared when the remaining supporting agriculture could no longer feed the large concentrations of people. This process is occurring in America now (22).

This happened repeatedly and is happening now, because agriculture has always been managed on a crisis basis of reacting to problems. We get very involved in problem solving. "Problem solving provides an almost automatic way of organizing your focus, actions, time and thought process. Problem solving can be very distracting while at the same time giving you the illusion that you are doing something important and needed" (23).

The problem with problem solving is that it is not an end in itself. When a problem is solved, something is taken away, creating a void that is filled by another problem. In contrast, when something is created, for example healthy farms and rural communities that will support families for generations, something is brought into the world. Problems should not be focused on as if they are goals. Without an appropriate and clear end goal of what we are trying to create, we can solve problems forever, but we won't accomplish anything meaningful and lasting (23).

Farm Family Quality Of Life

A sustainable rural culture depends not only on fixing what's wrong with conventional agriculture in terms of soil erosion, environmental pollution, and farm profitability, but also on improving the farm family's quality of life. One of the biggest problems with year-round confinement dairying, for example, is that it involves high capital investment in facilities and equipment and large amounts of purchased feed, fertilizer, and pesticides. Confinement dairying (originally called "zero grazing") with all of its required inputs supposedly was to solve the problem of pasture management by eliminating pasturing, thereby saving labor and increasing profits. But the result has actually been the opposite. None of the required "labor-saving" equipment operates by itself, and it breaks down.

Farmers feeding year-round in confinement work *a lot* more, have a lower profit margin, and consequently have a poor quality of life, compared to farmers who feed cows on well-managed pasture. Unnecessary year-round confinement feeding is good for businesses selling feed, equipment, and veterinary care, but not for farmers.

The work, debt load, and stress associated with high-input, high-capital confinement livestock production cause many people to quit farming, and also make farming less attractive to farm children, who look for other ways to make a living, usually in the cities. Fewer farmers on the land results in serious social problems of deteriorating rural communities and landscape, and high unemployment in cities.

As small farmers sell out to increasingly larger farmers, towns deteriorate because many smaller farmers spend more in town and support more services (e.g. churches, libraries, hospitals, schools, gas stations) than a few large farmers do. The overall rural population, number of retail stores and service enterprises, and school enrollment are a function of the *number* (therefore size) of farms in a local area, not the *volume* of farm goods produced in the area (25). In Vermont, for example, 57,000 jobs (17%) are directly linked to agriculture.

Another symptom of conventional farming problems

is that the divorce and suicide rate among farmers is one of the highest for any group (7).

The Rural Landscape

As farm families struggle to stay together and survive financially, land stewardship suffers and environmental degradation increases (13). This has resulted in another serious problem for all of society, as environmental pollution increases from high levels of pesticides and fertilizers applied to produce feed without adequate crop rotations, as is generally done to support year-round confinement feeding of livestock. Excess nutrients and pesticides leave the land in runoff, eroding soil, and leachate to pollute surface and ground waters. Consequently, the Environmental Protection Agency has identified agriculture as the largest nonpoint source of surface water pollution! Little or no crop rotations, plus no animals pastured on the land, plus applying excessive amounts of fertilizer and pesticides, results in sick farmland. Symptoms of sick farmland include unusually severe pest problems and the need to apply ever larger amounts of fertilizers and pesticides to maintain crop yields.

This lack of crop rotations and reliance on monoculture cropping of corn and other row crops is the most serious, though least obvious defect of the year-round confinement practice. Quietly and gradually soil organic matter and humus have been decreasing under monoculture cropping, until the amount of soil erosion occurring now equals or is worse than that of the dust-bowl years. Civilizations depend on farmers producing more food than the farmers need. Farming depends on the soil. Other civilizations have disappeared because they destroyed their soil base.

Toward A Permanent Solution

Agriculture, whether conventional or "sustainable",

currently is managed with no clear goal. So is everything else we do, including mining, logging, designing our cities, and the environmental movement. All of these activities involve groups of people in conflict with each other. If we would step back from these attitudes of conflict, we might see what we really want for America. More than likely everyone in all the various groups want similar things. For example, we want healthy land that's rich in species, cultures, and communities. We want satisfying work that gives us a sense of worth. We want loving relationships with our children, and we want their future to be good. We want a sense of tradition, without being trapped by the past. We want a dependable food supply for ourselves and generations to come.

Defining a goal enables people to know where they are headed. It takes a continuing effort to set aside old assumptions and think in new ways. This concept, of focusing on the end goal instead of on problems, is simple but involves a radical change in how we do things.

We must learn to live on what we have, and learn to live together. A healthy America requires a healthy rural culture, one that is revitalized both in population and economically more sound than what exists now. We need to champion rural culture by telling its stories, searching out practical information that can nourish it, and disseminate it as quickly as possible. We must do this; the alternative is intolerable (12).

Holistic Management
"... in studying our ecosystem and the many creatures inhabiting it, we cannot meaningfully isolate anything, let alone control the variables. The earth's atmosphere; its plant, animal, and human inhabitants; its oceans, plains, and forests; its ecological stability; and its promise for mankind can only be grasped by observing the dynamic interrelationships that constitute its being. Isolate any part, and neither what you have taken nor what you have

left behind remains what it was when all was one." We must work with wholes, rather than trying to fix a part in isolation from other parts (21).

Remaining family farms currently are unsustainable because of the following characteristics or practices, all of which must be changed to make farming sustainable:

- Farmers generally fail to consider the "whole" that they are managing. All too often they focus attention on aspects of the farm, such as corn or milk yields, to the detriment of the whole farm.
- The industrialization of agriculture has led to most farmers' decisions being based mainly on narrow economic views that treat farms as factories with separate parts. This kind of thinking ignores the farmer's main assets of productive soil, clean streams and rivers, genetic diversity, his/her health, and the family's stability.
- Leaving soil uncovered over winter or fallow for any period promotes soil erosion. The amount of soil lost *every* day from American farmland would fill a train of hopper cars 117 miles long!
- The separation or exclusion of animals from crop production has adversely affected the natural process of decay and humus formation.
- Inadequate goals prevent farmers from achieving anything meaningful and lasting.

To be sustainable, all farming activities must be oriented toward reaching a goal that takes into account the values of the people involved, what they need to produce to sustain those values, and the kind of landscape that will support everything that they want to accomplish (4).

Typically, farmers are besieged daily by salesmen, and advised by veterinarians, feed dealers, and Extension agents. If the farmer has no goal, he or she cannot know if the products being sold or the well-intentioned advice make positive contributions to the farm's well being. Consequently, nothing may be solved by the products or advice, and in worst cases they may finally drive the farmer into bankruptcy.

Farmers need a quick, simple way of critically assessing available products and advised practices, so they can select those that will help them move toward a well-defined goal. They need a method of integrated (holistic) management for the whole farm.

Fortunately, an integrated management method already exists to help farmers manage their farms so that they can become sustainable. Holistic Management (21) is a comprehensive thought model that helps people to see the effects of their actions on the complex processes operating in the ecosystem, and guides them to make decisions oriented toward reaching their goal of improved profitability and quality of life.

The Holistic Management model is relatively simple to use. It involves no specialized knowledge, but explanation and assistance by an experienced person is helpful during the first stages of its use. The model requires you to define a goal that includes three components:
• The quality of life you and your group desire, based on your values;
• What you will produce to achieve this quality of life;
• A description of the resource base as it has to be far into the future to sustain what you produce.

Only the farm family knows its reality. They are the experts, and the only ones that can set a goal and manage their whole farm accordingly. Once the goal is defined, and everyone on the farm is oriented toward achieving the goal, purchases and practices can be tested quickly along specific guidelines to ensure that the quality of life you seek is attained, and that all your decisions are truly sustainable--humanely, economically, and ecologically (23).

The model's testing guidelines are applied by asking relatively simple questions:
• Whole Ecosystem: Position of farm on environmental

brittleness scale? [All environments, regardless of total rainfall, can be classified on a scale from 1 (nonbrittle) to 10 (very brittle), ranging from jungle to true desert.] Will the proposed action move the ecosystem foundation toward or away from the landscape part of the goal?

- Weak Link (used in 1 or more of 3 situations):
1) Logjam: Is anything holding up steady progress to the goal?
2) Endangered or Problem Organisms: Are we looking first at the weakest part of the life cycle?
3) Solar Chain: Are we addressing the weakest link this year in the chain from solar energy to money received for products sold?
- Cause and Effect: Are we dealing with a problem or a symptom? Could this action unleash many problems (symptoms) by defying succession?
- Marginal Reaction (comparing 2 or more actions):
Which action will give the greatest movement toward the goal for each additional unit ($ or labor) of effort?
- Energy/Wealth Source/Use:
1) What is the source of the money/energy (e.g. external, internal, benign, interest)?
2) What is the pattern of use proposed (e.g. cyclical, consumptive, addictive, building infrastructure)?
- Gross Margin Analysis (to compare enterprises):
What does each enterprise, after covering added costs, contribute to covering overheads?
- Society & Culture (feelings more than thoughts):
Will this action really lead to the life we desire and what will it do to others? (21)

Management Intensive Grazing

One solution to the problems that farmers and rural communities are experiencing may lie in better use of pastureland. It is an enormous resource that, together with rangeland, covers 47 percent (887 million acres) of the total area of the continental USA.

Better use of the pasture resource helps solve the problems of low profitability and excessive work load, especially for dairy farmers. Feeding livestock on pasture can cost only 1/6th as much as feeding them in confinement. Benefits of pasture feeding, compared to confinement, accrue from:

- Decreased costs of grain and concentrate supplements, cropping, harvesting, storage, feeding, and manure handling;
- Less machinery use, repair, and fuel (25-40% less) needs;
- Lower hired labor requirements;
- Sale of surplus feed or less purchased feed;
- Lower veterinary costs, because animals on pasture are healthier and have better reproductive performance;
- Longer productive life for livestock;
- Premium price for high quality milk (low somatic cell count) from pastured cows.

Society may benefit in various ways from widespread livestock feeding on pasture and improved farm family quality of life. As livestock are fed more on pasture, farmers spend less on feed sold by outside corporations (this money drains rural communities), and spend more on products and services of local towns and villages, potentially revitalizing entire rural communities. Ripple effects generate $1.50-2.80 in a community for each $1 spent within the local economy (9). Because more farms may remain in business due to higher profitability and less work, and more farm children may go into farming because it is perceived as a desirable occupation again, the rural landscape will be maintained and rural communities will be rejuvenated.

Also, substituting pasture for corn and other tillage crops can significantly decrease soil erosion and environmental pollution from pesticides, fertilizers, and manure. The 80 million acres of corn grown annually in the USA receive 55% of herbicides (220 million lb), 44% of insecticides (9 million lb), and 44% of nitrogen fertilizer (5 million tons) applied each year (19). (see other environmental benefits below).

Management intensive grazing has proved to be a relatively simple way for farmers to use the existing largely wasted pasture resource in feeding livestock to cut production costs and reduce labor needs, thereby gaining the money, time, and energy needed for them to improve their family's quality of life.

Holistic management and management intensive grazing are compatible, and each actually needs the other for greatest benefit to the farmer.

What Needs To Be Done?

Although we know how to manage the pasture resource better than we did several years ago, we have only just begun to learn about managing pastures and livestock on pasture in the United States. Compared to other countries such as New Zealand, Ireland, Scotland, England, France, and Holland, America's investment and efforts in developing and using its huge pasture resource are extremely small. The potential for low-input, highly profitable, sustained livestock production on pasture is enormous.

To realize that potential, however, we must continue and increase our pasture and livestock management research and outreach efforts. These efforts must be guided by and respond to farmers who holistically manage their farms. Otherwise researchers easily can spend public-supported time, effort, and resources answering questions no one is asking, or answering the wrong questions posed by farmers who don't know how to manage their farms holistically. New Zealand agricultural researchers and Extension personnel are required to either have their own farms or work two days a month on a commercial farm. This is to keep their feet on the ground, and reduce the number of hair-brained recommendations to farmers.

The shift to low-capital, management-intensive,

highly profitable pasture-based dairying is occurring in Vermont and in other states. But the transition to the new practice may not be occurring quickly enough to save most of the remaining dairy farms. Discarding an old routine (paradigm) and putting a new one into practice is an emotionally and mentally painful process, and most people hesitate to do it, even if the old paradigm threatens their livelihood (17).

In addition, technical help is needed to enhance farmers' abilities to productively use a more management- intensive pasture-based system. Unless farmers receive the assistance and support they need to change to the new paradigm, many of the remaining dairy farmers in the United States probably will go out of business, with disastrous consequences for rural culture and the country. Unfortunately, just when more highly trained Extension agents are needed to help farmers make this change, funding of Extension activities in agriculture have been reduced.

To meet the increasing need for information and assistance in using the pasture resource, we formed a Grazier Support Network in Vermont. Individual frequent assistance and group meetings on farms to improve grazing management clearly is helping farmers participating in the Network to increase profitability and reduce labor needs. (see chapter 12 for further discussion of Grazier Support Networking)

ENVIRONMENTAL BENEFITS
One of the major goals of the Conservation Reserve Program (CRP) is to reduce water and wind erosion by establishing perennial grasses and legumes on more than 45 million acres of highly erodable and environmentally sensitive cropland that has been under cultivation. The benefits of perennial grassland in not only stopping soil erosion, but forming new soil and adding to soil organic matter are well known and documented. Besides stopping

soil erosion and building soil, grassland potentially can decrease atmospheric carbon dioxide levels. When grassland is used for feeding livestock under management intensive grazing, it also reduces the amounts of nitrogen, phosphorus, and pesticide entering and leaving farms.

Carbon dioxide

The level of atmospheric carbon dioxide was about 265 parts per million (ppm) in 1850, at the beginning of the industrial revolution. The level since then has increased to 340 ppm as a result of burning fossil fuels, clearing forests, and plowing grassland. If unchecked, the level could increase to 600 ppm during the next 50 years. Since carbon dioxide acts as an insulator, its increasing level in the atmosphere may be resulting in a steady warming of Earth. It is projected that the average temperature could rise 5°F (3°C), with a 13°F (7°C) increase at the poles within 50 years. This could have catastrophic worldwide climatic consequences that would affect agriculture and coastal populations.

If the amount of land in permanent grasslands would be increased enough, the level of atmospheric carbon dioxide could be reduced to preindustrial levels. This is because perennial grasses and legumes use carbon dioxide in photosynthesis, converting it to carbohydrates, which are used to build new plant material and are stored as food reserves in roots and rhizomes. As plants regrow in spring or after each grazing or cutting, root material separates from plants as carbohydrate root reserves are translocated to develop new shoots and leaves. After adequate leaf surface has developed, carbohydrate reserves are replenished in new roots and rhizomes. The amount of root material produced may be three times greater than shoot and leaf production.

Although unused, CRP grasslands are gaining an average of 1000 lb carbon/acre/year. These results suggest that the 45 million acres enrolled in CRP are recovering about 45% of the 84 billion pounds of carbon released annually from US agricultural activities (10).

The repeated pulsing of root formation and loss that occurs as plants are grazed and regrow under intensively managed grassland can remove considerably more carbon from the atmosphere and store it underground than is removed by unused grassland. For example, a permanent pasture intensively managed to yield 4 tons DM/acre/year, also produces about 12 tons/acre/year of carbon-rich root and rhizome material.

This translates into a significant amount of carbon dioxide that could be "permanently" removed from the atmosphere by converting cultivated land to productive pastures. Pasture soil contains about twice the organic matter of tilled soils. Due to this difference, about 78 tons of carbon dioxide could be removed from the atmosphere and stored in the soil in each acre of land converted from cultivated crops to pasture. (20)

Nitrogen, Phosphorus, Pesticides, And Soil Erosion

Surface and ground water quality has been degraded especially by soil sediments and nitrogen, phosphorus, and pesticide runoff and leaching from cultivated cropland (19). Farmers feeding livestock on intensively managed permanent grassland require and purchase less fertilizer, pesticides, and supplements. These input reductions could benefit surface and ground water quality, because less pollutants would enter and leave permanent grassland-based farms.

We estimate that for *each* dairy cow fed during 6 months on intensively managed pasture, the following reductions in farm inputs and activities occurs:
- 2394 lb less grain and concentrate supplement fed.
- This concentrate contains about 0.65% phosphorus, so this would result in 15.6 lb less phosphorus entering the farm.
- 135 lb less mineral supplement fed, containing 10%

phosphorus, resulting in 13.5 lb less phosphorus entering the farm.

- 0.1 acre less land needed in corn production, with consequent reductions of 15 lb less nitrogen fertilizer and 0.7 lb less herbicide applied.
- 8 lb less phosphorus/acre/year applied on permanent grassland-based farms, compared to conventional farms that rely on cultivated cropland, because of nutrient cycling that occurs with grazing livestock.
- Less soil erosion, which reaches about 90 ton/acre/year on sloping soils planted to corn.

These savings and benefits would be in addition to savings in fuel, machinery repair and replacement, labor, and veterinary costs due to improved herd health. Is it any wonder why there's resistance to farmers feeding livestock on pasture? Reducing farm input needs directly benefits farmers, society, and the environment, but not agribusiness. If these kinds of issues would be taken into account by economic enterprise analysis, *then* we could see how beneficial and profitable feeding dairy cows and other livestock on pasture really is, compared to unnecessary year-round confinement feeding.

ECONOMIC BENEFITS
Not so very long ago economic studies concerning the feeding of livestock on pasture under management intensive grazing in the United States were rare. Now they can be found more easily, ranging from individual farm case studies to linear programming models that apply to many farms. The studies invariably show that farmers who use Voisin management intensive grazing receive benefits that significantly exceed additional costs of adopting the method.

Costs include fencing as one of the first considerations when converting to management intensive grazing. Many farms already have perimeter fences in place, and only paddock subdivision are needed. Sometimes only portable fencing needs to be purchased. Since benefits and costs are compared on an annual basis, fencing costs are amortized over their estimated useful lives to convert to an annual cost basis. Costs of providing drinking water to grazing livestock also are amortized to convert to an annual basis. Costs of moving animals and portable fencing, and providing water are based on the amount of time needed to do the work. This labor is given an arbitrary value per hour to include it as a cost. In some situations additional costs may exist, such as seeding, fertilizing, or mowing.

Benefits of converting to management intensive grazing reflect the improved quantity and quality of forage produced from pasture. This results in purchasing less grain, concentrates, silage, and hay. Since less of these need to be fed, the amount of labor is reduced. It takes two to three times as much labor to feed the large amount of silage, hay, grain, and concentrates required by confined animals as it does to feed the reduced amount needed by animals under management intensive grazing.

Other benefits include less labor for manure handling and decreased bedding expense, because animals are confined less when grazing. Since less crops have to planted, fertilized, cared for in controlling pests, harvested, stored, and fed for animals under intensively managed grazing, less labor is needed in these activities. It follows that less machinery use and repair, fuel, seed, fertilizer, pesticides, and electrical power are needed with decreased cropping. With decreased cropping there would also be less soil erosion and water pollution, but our society still doesn't place an economic value on these, so they aren't included in economic analyses.

Other benefits may not be immediately quantifiable, such as improved herd management due to closer observation of animals when moving them in the pasture. Herd health generally improves. Both of these aspects can save veterinary expenses, and result in improved conception, production, and longevity of livestock. On dairy farms, milk quality may improve by grazing cows on well-managed pasture, and result in premiums received for the milk.

Because dairy farmers have been experiencing severe financial problems lately, most of the studies that have been done concern dairy cow feeding on pasture. All of the studies show the same thing: it pays to feed livestock on well-managed pasture as much and for as long as possible during the year. Some of the studies are presented here to prove the point.

DAIRY

Vermont Farms
Farmers (23 in 1994; 21 in 1995) who participated in the University of Vermont Grazier Support Network (see chapter 12) provided the production data (Table 11-1) used in this study comparing the economics of pasture-based dairying to confinement dairying in Vermont. The pasture-based dairy farmers earned an average of $600 net cash income per cow over the 2 years. In contrast, 24 dairy farmers, comprising the top 25% in per-cow profitability of farms using Agrifax accounting and presumed to be confinement dairies, only earned an average of $451 net cash income per cow (Table 11-2).

It should be noted that the pasture-based sample included any Network farmers who provided economic data. This included a wide range of farmer pasture management expertise, pasture sward development, and per-cow profitability, in contrast to the confinement sample that was limited to the 25% most profitable farms.

Table 11-1. Financial analyses of pasture-based dairy farms that participated in the University of Vermont Grazier Support Network.

Item	1994	1995	Average
Number of farms	23	21	22
Milking cows per farm	59	61	60
Total head of cattle	99	97	98
Milk production/cow, lb	15,700	15,812	15,756
Avg milk price/cwt, $	13.79	13.22	13.51
Cull rate, %	19.5	10.2	14.9
Total full-time workers	2	2	2
---------------------------per cow averages---------------------------			
Total acres	3.5	3.5	3.5
Open acres	2.5	2.4	2.5
Grazing acres	1.7	1.6	1.7
Income			
Milk sales, $	2165	2106	2137
Cattle sales, $	147	132	140
Crop sales, $	6	15	11
Other sales, $	50	45	48
Total income	2368	2308	2338
Expenses			
Marketing & hauling, $	166	180	173
Feed purchases, $	634	640	637
Vet & medicine, $	51	46	49
Breeding, $	28	33	31
Hired labor, $	130	126	128
Custom & machine hire, $	33	51	42
Records/DHIA/tax prep, $	29	23	26
Fertilizer, lime, pesticides, $	27	20	24
Seed, $	7	3	5
Repairs, $	101	102	102
Fuel/oil/grease, $	30	30	30
Bedding, $	28	20	24
Supplies, $	100	102	101
Utilities, $	73	78	76
Rent, $	75	86	81
Taxes, $	35	36	36
Insurance, $	41	47	44
Interest, $	134	126	130
Total cash expenses, $	1725	1716	1721
Net cash income, $	642	557	600

Table 11-2. Comparisons of Grazier Support Network (GSN) pasture-based dairies (23 in 1994; 21 in 1995), compared to 24 most profitable (Agrifax top 25%) confinement dairies, Vermont.

Item	GSN Pasture based 1994-95 avg	Top 25% Agrifax confinement 1994*
Number of farms	22	24
Milking cows/farm	60	155
Total head of cattle	98	262
Milk production/cow, lb	15,756	18,788
Avg milk price/cwt, $	13.51	13.87
Cull rate, %	15	not available
Total full-time workers	2	3.3
Open acres/cow	2.5	2.1
	1994-95 avg	
Interest, $	130	256
Hired labor, $	128	330
Vet & medicine/cow, $	49	not available
Purchased feed, $	637	733
Fertilizer/lime/pesticides, $	24	85
Seed, $	5	24
Repairs, $	102	162
Fuel/oil/grease, $	30	57
Total cash income/cow, $	2308	2736
Total cash expenses/cow, $	1721	2285
Net cash income/cow, $	600	451

*A complete financial analysis was not available from Agrifax in 1995.

Almost everyone in the pasture-based sample fed too much protein supplement. As their management ability increases, their pasture forage quality improves, and they select cows that perform better on high-quality pasture forage, they will be able to feed less supplement, thereby improving their farm profitability even more. The contrast in profitability between pasture-based and confinement dairying likely will become greater, in favor of pasture-based dairying.

This study was done by Jon Winsten, assisted by Sarah

Flack, Dave Hoke, Lisa McCrory, Joshua Silman, and the farmers who took the time and made the effort to provide accurate, dependable financial information. Jon put everything on a per-cow basis to make it easier for you to make direct comparisons with your farm.

The Opitz Farm

At the time of this study in 1989, Charlie Opitz fed 800 Holstein milking cows and 1200 dry cows and heifers on 2000 acres near Mineral Point, Wisconsin. He produced 12- to 13-million pounds of milk per year. (Charlie now feeds more than 1200 cows on pasture!) During 7 to 8 months of the year the cows and heifers get most of their feed by grazing under intensive management. In 1989 Charlie calculated that his farm produced about $800 worth of milk per acre. Some of the better pastureland returned up to $1300 per acre to management and labor, after costs for seed, fertilizer, purchased feed were deducted!

The decision to pasture the milking cows came in 1987, when Charlie was dealing with heat-stress problems in the confinement barn. He put 80 milkers on pasture under management intensive grazing in the spring of that year just to see what would happen. During one hot spell, milk production from cows in the barn dropped 22 percent, while production from the grazing cows fell only 12 percent. The 80 grazing cows made $900 worth of milk per acre above supplemental feed costs in 6 months grazing! That was enough for Charlie; now all cows graze.

Charlie prepares for spring grazing beginning in the previous autumn. He stops grazing about two-thirds of his pastures in early September. Half of the deferred pasture is carefully rationed out to heifers and dry cows during the fall, winter, and early spring. In the spring, forage on the deferred pasture still contains 14 to 18 percent crude protein! Deferring pasture from fall to spring enables Charlie to begin grazing early on well-drained soil.

Milking cows start grazing in mid-April and stop grazing sometime in November, depending on the season. Starting the first rotation early in spring is essential to get staggered regrowth for the whole season. Milking cows graze a paddock first for 1 to 2 days, followed by dry cows and heifers for 3 to 6 days. Recovery periods between grazings vary, depending on plant regrowth and pasture mass (6 inches tall). Heifers are also used to clean up hay fields right after the hay is removed, thereby gaining a day or two in recovery time for the pasture.

The rolling herd average milk production is 14,000 to 15,000 pounds. Before putting the cows on pasture, the herd average was 17,000 pounds when cows were milked three times a day. Feed and other cost savings have more than made up for the decrease in milk production. Now that cows are being milked twice a day, only six or seven people are needed to milk, compared to the 12 needed before.

During the grazing season, half of the cows' feed comes from pasture, and half is fed in the barn at milking. The ration for milkers consists of 6 to 7 pounds of alfalfa hay, 1 to 2 pounds of sudax or small grain (wheat) silage, and 12 pounds of concentrate containing varying combinations of wheat middlings, hominy, wet gluten, distiller's grain, and full-fat soybeans. In winter the concentrate portion is increased to 24 to 27 pounds.

Charlie double-crops the sudax and wheat, to produce about 40 tons of silage per acre. Analyses showed that the pasture forage averaged 22% dry matter, 22% crude protein, 24% acid detergent fiber, and 40% neutral detergent fiber during the 7-month grazing season.

Charlie estimated that it cost $40 to $70 per acre to install fencing and drinking water. He says that's a bargain, considering that it costs about $20 a trip to cut silage or $100 per acre to make hay each year. He feels that farmers can pay off their land debt, just with the savings that they can make in operating costs from using controlled grazing management. Annual fuel consumption on his farm is now $30/cow, compared to the Wisconsin state average of $60/cow. He used to cut about 3000 acres (including rented land) of hayland each year. Using management intensive grazing has reduced the need for hay so much that now he only cuts 1200 acres of hayland. Following his estimate of how much it costs to make hay per acre, Charlie makes a gross saving of $180,000 a year (3000 - 1200 = 1800 acres x $100) on reduced haying costs. No wonder he smiles a lot!

As a result of his grazing management, Charlie's pastures now contain mostly quackgrass, bromegrass, tall fescue, orchardgrass, bluegrass, alfalfa, red clover, and birdsfoot trefoil. He feels that there is no optimum grass or legume for the entire farm. A combination of plant species is needed to cover the shortcomings of each.

He has come to prefer bromegrass in his pasture. If it has nitrogen from associated legumes or manure, bromegrass contains up to 21% crude protein, and can provide five grazings a year, carrying nearly 2 cows per acre. Quackgrass also performs very well for Charlie. He grazes it five times on high fertility soil, and it often contains more than 30% crude protein!

More fun and profit aren't the only benefits that Charlie has realized since he went to management intensive grazing. Local Soil Conservation Service officials estimate that Charlie's land was losing about 90 tons of soil per acre per year before he took the slopes out of corn production and seeded them to perennial forages. This not only stopped erosion and silting, but also almost entirely eliminated pesticide applications. Charlie says that grass farming solves 99% of the problems that the USDA's Sustainable Agriculture Research and Extension program is trying to deal with. Once the source of problems (annual row cropping, especially corn) is removed, symptoms disappear.

Charlie Opitz's farm is proof that Voisin management intensive grazing isn't just for small herds, but is a profitable alternative for any size dairy farm (5, 14).

Conclusion

These results lay to rest once and for all the misconception that pastures in the United States cannot be used to feed dairy cows profitably. The misconception resulted from the mismanagement of pastures in the USA, not from anything inherently bad about pastures.

The alternative paradigm is that there are at least two ways of feeding dairy cows: on intensively managed pasture or in year-round confinement. Both ways need and deserve research and Extension effort. Until the alternative paradigm of feeding dairy cows on pasture was accepted, farmers could not consider doing it, university researchers could not work on problems related to it, and Extension agents had only misinformation about it to extend.

The year-round confinement feeding practice has received overwhelming attention by university researchers funded by agribusinesses that had a vested interest in continual increase of farmers' purchases of production inputs. Widespread feeding of livestock on well-managed pasture benefits farmers, society, and the environment, but does not directly or sufficiently benefit individual agribusinesses, since it aims at reducing off-farm purchases of inputs. Therefore, funding of research to reduce farm production inputs by feeding livestock on pasture has to come from government, foundations, or farmers.

SEASONAL DAIRYING

Once you accept that there can be at least two ways of feeding dairy cows, it's easy to accept the idea that there are at least two ways of milking them. A herd can be milked year-round so that you have continual work to do. (Why you'd want to do that, I don't know, but, "To each her own," said the woman as she kissed the cow.) Or a herd can be milked seasonally to take advantage of something, such as cheap feed (e.g. intensively managed pasture) or higher prices (e.g. fall premiums).

Least-cost dairying results from feeding cows on intensively managed pasture for as long as possible during the lactation. That requires seasonal dairying. If we as a nation can get organized on this, everyone can win with seasonal dairying.

Two thirds of the milk produced in the USA is made into manufactured products. In temperate regions it makes no sense at all to produce milk year-round, half the time on stored feed, to make those products. If farmers in the North changed to seasonal dairying on pasture, it's true that cheese plants would have to close down for 2 months a year. But this kind of seasonality is common. For example, cotton gins and sugar mills close for 8 to 9 months a year.

Winter fluid milk needs of the entire country can be supplied with only 1/9 of the herds lactating. Farmers in the Southern states and California could produce this small amount of milk seasonally on their excellent winter pastures.

Surely some way could be found so that seasonal dairy farmers could get paid year-round. For example, in Louisiana sugar cane producers are paid for their cane by the sugar mill cooperatives over 12-months, based on the average price of sugar for each month. The farmer-owners of milk cooperatives should be able to develop a similar plan for themselves (18).

Ohio State University

Dr. David Zartman, Head of Ohio State University's Dairy Science Department completed a 5-year study on seasonal dairying (spring-calving and producing most milk on pasture) in 1991. His findings showed that by feeding 60 spring-calving cows on pasture, a farmer would net about $36000 per year, plus have a vacation from milking for 2 months during the winter when the entire herd is dry!

Milking cows were allowed just enough pasture forage so that they consumed about 60 percent of the available forage in a 24-hour grazing period. Then they were given a fresh paddock, and heifers followed behind grazing the remaining forage. Recovery periods ranged from about 15 to 30 days, depending on plant regrowth. In the spring, surplus forage was harvested as hay for winter feeding.

The key to seasonal dairying is synchronizing milk production with pasture forage availability. The cows should be at their greatest nutritive requirement when pasture plants are growing the fastest, and the forage has the highest protein and energy levels. According to Dr. Zartman, farmers have been incorrectly sold the monoculture farm management model that includes feeding, breeding, and managing cows in the same way to get maximum milk production out of them no matter where they are. All evidence indicates that the model is invalid. He says that farmers need to try alternatives that fit local environments and economic needs.

At the experiment farm near Youngstown, cows received 1 pound of 14% protein supplement per 5 pounds of milk produced, or a minimum of 10 pounds of supplement per day. Dr. Zartman kept supplemental energy levels a little short, so that the rumen micro-organisms would do a better job digesting forage. He says that production was somewhat less, but it was more efficient economically.

In the first year of the study, commercial grade Holsteins averaged 15,154 pounds of milk, and Jerseys averaged 11,353 pounds. During the second year, pasture quality had improved, and milk production increased to 17,865 and 12,458 pounds.

Cows and heifers were synchronized for breeding with Synchromate-B implants. They were bred to begin calving on March 10. Any animals that bred late and weren't due to freshen by May 10 were sold. The herd was dried off on

December 20, after an average lactation of 270 days.

One problem with seasonal dairying is that all of the calves arrive within a short period. But at that time there are no crops to worry about, and once grazing begins barn chore demands decrease greatly. The other problem is that the price of milk is lowest when most of the milk is shipped. But that's acceptable because the milk is produced at a much lower cost on pasture than it is in confinement.

Benefits include improved herd health, with little or no mastitis. Cows aren't subjected to the stress of calving in winter, and frozen teats are no longer a problem. The best part of seasonal dairying on pasture is its low capital requirement. All that's needed is a barn, well-managed pasture, a manure spreader, and a tractor. Supplements, hay, and silage can be purchased, or forage can be custom-harvested for storage as hay or silage.

Dr. Zartman developed a computer spreadsheet that farmers can use to analyze their records and project how changes in management or pricing will affect their profitability. For more information write to Dr. Zartman at the Dairy Science Department, Ohio State University, Columbus, OH 43210-1094 (6).

University of Wisconsin

Larry Tranel, a University of Wisconsin Extension Agent, who specializes in farm financial management, compared pasture-based seasonal to conventional confinement dairying. Based on 50-cow herds, he found that seasonal dairying on intensively managed pasture provided a profit advantage of $284/cow and $2.92/cwt of milk over confinement dairying! Seasonal dairying also enabled the farmer to take a 2- to 3-month vacation from milking, whatever that's worth.

Larry's analysis (Table 11-3) considered two 50-cow herds, each on 160 acres of productive land in Wisconsin. A spring-freshening, seasonal dairy with a 14,000-lb herd average and based on pasture, was compared to a year-

TABLE 11-3. Economic comparison of a spring-freshening, pasture-based seasonal dairy, to a year-round confinement dairy in Wisconsin.

Item	Seasonal	Confine-ment	Seasonal	Confine-ment
Rolling herd avg, lb/cow/year	14,000	20,000	14,000	20,000
	$/cow	$/cow	$/cwt	$/cwt
Income				
Milk sales	1,631	2,400	11.65	12.00
Cull cow sales	162	193	1.16	0.97
Calf sales @ $125	119	119	0.85	0.59
Total income	$1,912	$2,712	$13.66	$13.56
Feed Expenses				
Supplements	44	411	0.32	2.05
Corn @$2.15/bushel	126	220	0.90	1.10
Pasture @ $30/ton	146	0	1.04	0
Hay @ $75/ton	243	537	1.74	2.69
Subtotal	$559	$1,168	$4.00	$5.84
Other Expenses				
Replacement cost	225	350	1.61	1.75
Labor & management	200	300	1.43	1.50
Interest on cows @ 8.5%	85	102	0.61	0.51
Manure handling	20	100	0.14	0.50
Vet & medicine	30	70	0.21	0.35
Supplies, breeding fees, DHIA, utilities	82	155	0.59	0.78
Bedding, other	10	50	0.07	0.25
Subtotal	$652	$1,127	$4.66	$5.64
Total Expenses	$1,211	$2,295	$8.66	$11.48
Profit	$701	$417	$5.00	$2.08

Notes: Cwt is per hundredweight equivalent; includes income from sale of cull cows and calves. Cull sales: seasonal $650/cow, 25% culling rate; confinement $550/cow, 35% culling rate. Corn feeding: seasonal 58.5 bushel/cow; confinement 102 bushel/cow. Pasture forage: 4.9 ton/cow. Hay needs: seasonal 3.2 ton/cow; confinement 7.2 ton/cow. Replacement cost: seasonal $900/cow; confinement $1000/cow. Interest: seasonal $1000 invested/cow; confinement $1200 invested/cow. (C. Shirley. 1993. Milking for money or for profit? *The New Farm*. Sept./Oct. p.33.)

round confinement dairy with a 20,000-lb herd average. Farm assets totaled $300,000 on both farms.

He assumed that the seasonal farmer had at least 5 year's experience with management intensive grazing, and at least 2 year's experience with seasonal dairying. Pasture forage quality at least equaled stored forage quality. Autumn stockpiling pasture forage enabled cows to graze 240 days, and obtain 60% of their dry matter need from pasture. Each cow produced milk for 220 to 270 days, and the entire herd was dry 2 to 3 months a year (24).

(Larry Tranel's Seasonal Dairy IBM-compatible, computer software program is available for $149 from: The Stockman Grass Farmer, P.O. Box 9607, Jackson, MS 39286-9607; Phone: 800-748-9808.)

Maximum Profits From Seasonal Dairying

Once you're comfortable with other ways of feeding and milking cows, then any other possible aspect of dairying should be considered on its merits without preconceived notions getting in the way. Right? OK, how about the possibility that it's as profitable to seasonally milk two cows producing 11,000 lb of milk/cow on well-managed pasture, as it is to milk one cow producing 22,000 lb milk in conventional confinement? Larry Tranel's analysis has shown that this is true. Although cows producing in confinement net more money per cow, milk from seasonally milked cows on pasture is cheaper to produce.

If the seasonally milked cows only produce 11,000 lb milk/cow, 100 cows would be needed to provide the same net income as 50 cows fed in confinement. But if the seasonally milked cows produce 14,500 lb milk/cow, only 50 cows would be needed to equal the net income from confined cows producing at 22,000 lb milk/cow. It's relatively easy to achieve a 14,500 lb herd average on intensively managed pasture. If you could handle 100 seasonally milked cows yielding 14,500 lb milk/cow, you could realize a return on assets of about 23%! (27)

If you'd like to determine which way of milking would be best for you, get Larry's Seasonal Dairy computer program. You can enter any numbers you want to calculate possible incomes, expenses, production levels, prices, asset levels, number of cows, cull rates, replacement costs, and dry matter intake from pasture.

The Patenaude Farm
Dan and Jeanne Patenaude farm in the hills of south-western Wisconsin. Nestled into the base of a forested ridge, their farm lacks many of the things generally associated with conventional dairy farms. There are no silos and no manure storage pit, no large shed filled with big tractors and machinery, no plowed land for growing corn or grain, and no debt. Instead there is a small well-maintained barn and a small shed for storing the tractor, haybine, and baler.

The other things that make this farm different from conventional dairy farms is its low labor requirement, high profitability, and synchronized breeding for spring calving to produce most of the milk on pasture.

The farm has 73 acres, with only 27 of those being tillable. There are 20 acres of permanent pasture, and the rest is woodland. A trout creek flows through fertile bottomland that used to grow corn, but now is covered with lush grass-red and white clover pasture.

The 24 high-producing (17000 lb average) Holsteins and young stock graze for about 220 days (April 15-Nov. 20). Dan gives the cows a fresh paddock after one or two milkings. Dry cows and heifers follow the cows to clean up remaining forage. Analyses showed that the pasture forage contained 20 to 26% crude protein and 0.70 Mcal NEL/lb DM.

Dan estimated feed costs to be $2.76/cwt of milk during the grazing season, calculated from a grain cost of $1.67/cwt and a forage cost of $1.09/cwt. This is about $4/cwt of milk lower than winter feed costs, and directly reflects the use of inexpensive pasture forage.

Because there were no satisfactory fencing materials available in Wisconsin when Dan and Jeanne fenced their paddocks, Dan became a dealer of New Zealand style fencing equipment. Since it takes so little of his time to care for his cows during the grazing season, Dan now has time to help other farmers design pasture layouts and set up fencing.

If it costs so much less to feed cows during the grazing season, why not produce most of the milk during that time, instead of working like a maniac to do it during the winter? When Dan asked himself that question, the next step was obvious. In autumn 1990 he sold all of his fall-calving cows, keeping or buying replacements that would begin calving in spring. He now dries the whole herd off just before Christmas and takes a well-deserved and needed 4-month vacation from milking (14).

BEEF

Cow/Calf Extended Grazing

Extended grazing usually involves grazing hay fields (meadows) during times of the year when hay otherwise would be fed to a herd. The longer that grazing lasts during a season, the less forage must be mechanically harvested and stored for winter feeding. Making maximum use of pasture by extending the grazing season reduces production costs and increases profitability. This 3-year study was done by West Virginia University researchers to determine which meadow management practice gives the most profit.

Four management practices were evaluated:
1) two cuttings of hay, no grazing;
2) one hay cutting, late fall grazing;
3) early spring grazing, two hay cuttings; and
4) early spring grazing, one hay cutting, and late fall grazing. The practices were compared for orchardgrass and

tall fescue.

Total forage production for both orchardgrass and tall fescue (hay + grazed) was highest (3.4 ton DM/acre) when meadows were cut once and grazed in late fall. Forage production was lowest (2.9 ton DM/acre) when meadows were grazed in early spring and cut twice. Tall fescue consistently produced more forage for haying and grazing than orchardgrass.

With the exception of costs on a per ton of hay basis, where the costs of cutting once and grazing in fall were lowest, the least-cost practice was spring grazing, one cutting, and late fall grazing. Spring grazing followed by two cuttings had the highest costs on a per ton of hay basis. But two hay cuttings with no grazing was the highest cost method, regarding all other aspects, such as costs per acre, per ton of dry matter, and per pound of calf produced. Except for costs on a per acre basis, average production costs were lower for farms using tall fescue than those with orchardgrass. Production costs were lower when hay was baled in 1500-lb round bales, than when baled in 50-lb square bales.

The practice of haying and grazing once resulted in the most calves sold. This is because more forage was produced with this method. The most profit in terms of returns to fixed resources, however, was obtained when meadows were grazed in spring, cut once, and then grazed in fall, for both orchardgrass and tall fescue. The least profit resulted when meadows were hayed twice and not grazed (tall fescue), or grazed in spring and hayed twice (orchardgrass).

The researchers concluded that early spring grazing, one hay cutting, and fall grazing is the most profitable way of extending the grazing season for cow/calf production. They also found that tall fescue generally performed better than orchardgrass in terms of forage production, cost, and profitability (8).

SHEEP

Since providing feed is the largest single cost in raising sheep (and all other livestock), it's the aspect where most can be gained by improving the feed supply (2,15). Whatever your reason for raising sheep, costs involved can be reduced greatly by feeding them on well-managed pasture as much as possible, as the following studies show.

The Wood Farm

Peter and Hilary Wood make a living entirely from 250 ewes on a 160-acre (70 tillable, 50 pasture, 40 woodland) farm in Wisconsin. By paying close attention to breeding and marketing, achieving nearly a 300% lamb crop, and grazing from April 15 to January 15, they net $82.20 per ewe per year (Table 11-4)!

The Woods have distinguished between terminal sire breeds, which produce terminal crosses only, and ewe breeds. They use Finn-Rambouillet cross ewes that have multiple births and good mothering characteristics. Their ewe lambs are in high demand as replacements, and all are sold before birth at a premium price.

By lambing in April the Woods take advantage of two periods of high quality and quantity pasture production. Lambs are born and grow on spring pasture; ewes are bred back on fall pasture. At lambing their animal numbers jump from 250 to 900, coinciding perfectly with the period of most pasture forage production.

Lambs are weaned selectively as they reach 30 lb liveweight in June; then they graze ahead of ewes until September. Wether lambs are sold as feeders whenever the pasture supply gets short, or on September 1, when ewe and ram lambs are sold. Between September 1 and October 20, ewes and replacement ewe lambs graze woodland (plus hay), while forage stockpiles on the pasture for fall and winter grazing. Ewes are bred during November 7 to 27.

Costs are kept low by:
- grazing sheep on well-managed pasture (orchardgrass, reed canarygrass, Kentucky bluegrass, timothy, birdsfoot trefoil) for at least 9 months;
- feeding harvested forage for less than 100 days;
- feeding concentrates (1/2 lb/ewe/day) only from mid-February until lambing;
- keeping the animals outdoors year-round; and
- controlling parasites (28, 29).

TABLE 11-4. Economics of sheep production on the Wood farm in Wisconsin in 1990.

Income per ewe	
Sale of wool:	
8 lb/ewe @ $1.30, including wool support	10.40
Sale of ewe lambs:	
1.35 lambs less 10% = 1.22 x $90	109.80
Sale of wether lambs: 1.30 x $40	52.00
Sale of ram lambs: 0.05 x $250	12.50
Total income per ewe	$184.70
Expenses per ewe, including lambs to market:	
Repairs/maintenance	9.50
Office	3.20
Veterinary	5.30
Fuel/oil/mileage	6.70
Taxes	9.40
Fertilizer	10.50
Insurance	6.40
Utilities	3.20
Labor/family	6.20
Feed (purchased)	15.30
Facility & equipment depreciation	14.00
Total expenses per ewe	$102.50
Net return to labor/management per ewe	$82.20

The Blazek Farm

Michael and Terri Blazek raise a small flock of Suffolks on a farm near Waldoboro, Maine. Their main objective with the sheep is to produce high value breeding stock for sale. They converted to Voisin management intensive grazing in 1985, after raising sheep for 8 years with conventional practices. Before they began managing grazing they used the land for producing hay, with some aftermath grazing.

This economic analysis was based on a "before" versus "after" approach. In each case the Blazeks used the same 10 acres of land. Before they began managing grazing, very little forage was obtained from the land. Eleven ewes weighing 200 pounds each grazed for about 45 days per season; 320 bales of poor quality hay was also harvested from the 10 acres. Additional needed feed was purchased.

In 1985 the Blazeks seeded 1 acre to Grimalda perennial ryegrass, and began managing the grazing of their sheep late that spring. They built a lane down the middle of their land, dividing it into two 5-acre areas. They placed black plastic tubing with four spigots in the lane to provide drinking water with a portable tub to paddocks. They formed twelve, 3/4-acre paddocks with portable fencing that they could remove to make hay if necessary. Three acres of land continued to be used mainly to produce hay.

Management intensive grazing improved the quality and quantity of forage produced on the farm. Since most of the land was used for pasture, the amount of hay harvested declined to 160 bales. But now 22 ewes (200-lb each) graze for 200 days, 9 lambs (135-lb each) graze for 45 days, and 2 rams (300-lb each) graze for 107 days. This totals 999 animal unit days.

(An animal unit day is the amount of feed needed to support 1000 pounds of animal weight for one day. For example, one animal unit day is the amount of feed needed per day for 8 ewes weighing 125 pounds each.)

To estimate the value of increased production, the additional animal unit days were valued in terms of their hay equivalent. The total cost of feed (pasture, hay, grain) needed before and after applying management intensive grazing was determined and divided by the number of animal units supported in each situation. The cost of feeding one animal unit decreased from $338 without managed grazing, to $170 with managed grazing. With 200-lb ewes, this meant that annual feed cost was reduced 50 percent, from $68 down to $34 per ewe. This cost savings was multiplied by the number of animal units that the Blazeks now can support, to obtain the increase in feed value of their land under management intensive grazing. The increased value amounted to $1042 per year.

Because the sheep were outside on pasture more under managed grazing, time needed for manual manure handling decreased by 51 hours. Valued at $5 per hour, this saved $255. For the same reasons, bedding expense decreased by $200, and time needed to feed hay and grain was reduced by 100 hours, for a saving of $500.

Costs of converting to managed grazing included fencing, the water supply, seeding one acre, moving animals every 4 days, and fertilizing the area where hay was cut.

Management intensive grazing proved to be very beneficial to the Blazeks. They estimated that they received about $2.60 in benefits for each $1.00 invested in managed grazing: (1917 ÷ 740 = 2.6). The economic advantages will continue to improve, because they intend to increase the number of animals carried on the 10 acres. So this analysis is conservative (Tables 11-5, 6).

Although not quantifiable, benefits were gained in other areas besides feed and labor savings. Michael and Terri feel that management intensive grazing enables their sheep to stay in better condition, and especially helps the ewes to recover faster from lambing. They feel that the quality of the pasture has improved drastically, with much higher white clover content, and the quality continues to improve. Since they handle and observe the animals more in moving them among paddocks, their livestock management has improved. This is especially valuable to them, because of their breeding stock operation (1).

TABLE 11-5. Economics of converting to management intensive grazing from conventional practices on the Blazek sheep farm in Maine.

1. Additional costs of managed grazing
   - Fence (amortized to annual equivalents)
     Permanent   290
     Portable   40
   - Seed (1 acre)   58
   - Water system (amortized)   27
   - Moving animals (includes providing water)   65
   - Fertilizer (machinery borrowed, no cost)   404
                        Total annual costs   884
                        Lower haying costs   -144
                                                                          $740

2. Benefits received from managed grazing
   - Improved feed value of 10 acres   1042
   - Reduced manure handling   255
   - Reduced bedding expense   200
   - Reduced labor in feeding   500
                        Total annual benefits   1997
                        Less hay produced   - 80
                                                                        $1917

3. Comparison of costs versus benefits of managed grazing
                        Annual benefits   1917
                        Annual costs   740
                        Net annual benefits $1177

TABLE 11-6. Comparison of feed per animal unit (AU) by source of feed on the Blazek farm before and after converting to management intensive grazing.

Pasture
   Before $54 ÷ 3.31 AU = $16/AU
   After $549 ÷ 6.22 AU = $88/AU
Hay
   Before $825 ÷ 3.31 AU = $249/AU
   After $330 ÷ 6.22 AU = $53/AU
Grain
   Before $240 ÷ 3.31 AU = $73/AU
   After $180 ÷ 6.22 AU = $29/AU

REFERENCES

1.Burns, P. and C.R. Jones. 1987. *Voisin rational grazing - sheep (Maine economic case study)*. Soil Conservation Service, Orono, Maine. Mimeo.

2.Burch, C. 1984. Dairy sheep update. *Sheep!*. 5(3):18-19.

3.Butterfield, J. and A. Savory, 1992. Sustaining Civilization. *Holistic Resource Management Newsletter*. Center for HRM. Albuquerque, New Mexico. 36:1.

4.Butterfield, J. & A. Savory. 1992. What Makes Farming Unsustainable. *Holistic Resource Management Newsletter*. Center for HRM. Albuquerque, NM. 36:2-3.

5.Cramer, C. 1990. Grass farming beats corn! -- The Charles Opitz Wisconsin dairy farm. *The New Farm*. Sept./Oct.   p. 10 - 16.

6.Cramer, C. 1990. Milk 10 months, take a vacation...net $36000! (Ohio State Univ. Seasonal Dairying) *The New Farm*. May/June. p. 12-14.

7.Davidson, O.G. 1990. Broken Heartland. The Rise of America's Rural Ghetto. Free Press, NY. 206 pp.

8.D'Souza, G.E., E.W. Maxwell, W.B. Bryan, and E.C. Prigge. 1990. Economic impacts of extended grazing systems. *American Journal Alternative Agriculture*. 5(3): 120-125.

9.Gable, D. 1986. Where our food comes from: the Hendrix College food system. Meadowcreek Project, Fox, Arkansas 72051. 48-minute video.

10.Gebhart, D. 1993. Some agricultural activities are related to carbon dioxide emissions. Kerr Center for Sustainable Agriculture Newsletter. 19(8):1.

11.Jordan, R.M., and W.J. Boylan. 1986. Sheep milk, cheese and yogurt production. Proceedings, "Adapt 100". *Successful Farming* . December.

12.Jordan, T. 1992. Beyond Conflict. *Holistic Resource Management Newsletter*. Center for HRM. Albuquerque, New Mexico. 37:8-10.

13. Lal, R., J. Kimble, and B.A. Stewart. 1995. World soils as a source or sink for radiatively-active gases. In R. Lal, J. Kimble, E. Levine, and B. A. Stewart (eds.) Soil Management & Greenhouse Effect. CRC Lewis Pub., Boca Raton, Florida. pp.1-7.

14. Liebhardt, W.C. 1993. *The Dairy Debate: Consequences of Bovine Growth Hormone and Rotational Grazing Technologies.* Univ. of California Sustainable Agriculture Research and Education Program.

15. Mitchell, D. 1982. *So You Want To Raise Sheep In Vermont....* Vermont Sheep Breeder's Association. Handbook. 48 p.

16. Nation, A. 1992. Spring lambing ideal for low cost pasture program. *Stockman Grass Farmer.* 49(2):1,5.

17. Nation, A. 1992a. Human resistance to change. Allan's Observations. *Stockman Grass Farmer.* 49(10):1.

18. Nation, A. 1992b. Seasonal milk production, in Allan's Observations. *Stockman Grass Farmer.* 49(12):21.

19. Natioinal Research Council. 1989. *Alternative Agriculture.* National Academy Press. Washington, D.C.

20. Rayburn, E.B. 1993. Potential ecological and environmental effects of pasture and BGH technology. In W.C. Liebhardt (ed.) *The Dairy Debate.* University of California, Davis. p. 270-271.

21. Savory, A. 1988. *Holistic Resource Management.* Island Press. 353 p.

22. Savory, A. 1992. Cities and Agriculture. Holistic Resource Management Newsletter. Center for HRM. Albuquerque, New Mexico. 36:7-8.

23. Savory, A. 1992. Beyond Conflict. Holistic Resource Management Newsletter. Center for HRM. Albuquerque, New Mexico. 37:9.

24. Shirley, C. 1993. Milking for money or for profit? *The New Farm.* Sept./Oct. p. 31-34.

25. Strange, M. 1988. *Family Farming: A New Economic Vision.* University of Nebraska, Lincoln. 311 pp.

26.Tranel, L.F. 1993. Maximum profits in a seasonal dairy. *Stockman Grass Farmer.* 50(9):16-17.

27.Tranel, L.F. 1993. Return on assets means grazing profits. *Stockman Grass Farmer.* 50(9):17-18.

28.Wood, P. 1990. Pastures, percentages, and profits: making a living entirely from sheep on a small farm. *The   Shepherd.* March. p. 8-13.

29.Wood, P. 1991. Matching the needs of the productive ewe to the seasonality of pasture: towards a least cost system of sheep production. *The   Shepherd.* January. p. 10-15.

12
Grazier Support

Treat others
As you would have others
Treat you.

In 1988 Dr. Ray Brougham, a distinguished New Zealand
pasture researcher, visited Vermont to speak at a pasture
management workshop. He told me that grazier
discussion groups had been absolutely essential for New
Zealand farmers to achieve least-cost (money and labor)
feeding of livestock on pasture, and sustain their rural
culture. That idea prompted me to find a way of forming
discussion groups in Vermont to accomplish the same
objectives.

I include this description of our program here to
suggest that you work to form discussion groups and a
grazier support network in your area. I believe that this
can be a first step in rekindling the sense of community
that almost has been lost in rural America. We must all
work together to help farmers stay in business and sustain
rural culture.

To apply research results, they must be transferred to
farmers in practical formats that fit easily into farming
operations. In the case of new pasture and livestock
management and economics information, special efforts
are needed to inform not only farmers, but also feed
dealers, veterinarians, and farm loan officers about the
potential benefits of feeding livestock on well-managed
pasture. This is because of past unfavorable experience
with pasture feeding that American farmers had due to
poor management practices.

In April 1992 we finally were able to begin a grazier
support program in Vermont, funded by the Kellogg
Foundation. We formed two discussion groups, each with
12 dairy farmers. In 1993 with funding from the USDA
Sustainable Agriculture Research and Education program

and increased funding from Kellogg, we were able to expand the program and include other livestock farmers.

We directly assist farmers to use pasture to its fullest potential in feeding cows and other livestock, to help the farmers achieve optimum production and profitability, and a better quality of life. Achieving these goals should help farmers to remain in business, and encourage their children to go into farming as a valuable and enjoyable career. Both of these accomplishments are needed to sustain agricultural production and revitalize rural communities.

During the grazing seasons of 1992-1995 while funding lasted, grazing management advisors visited each farm at least every 21 days, walked the pasture with the farmer and mutually learned all aspects of grazing management and feeding cows and other livestock on pasture. During those 4 years we worked directly with a total of 120 farmers, and we published a monthly newsletter distributed to 400 people.

Besides the frequent farm visits by advisors, once a month during the grazing season each of up to six discussion groups (including farmers, Extension agents, the advisor, Natural Resource Conservation Service personnel, and any other interested people) met for a pasture walk on a different group member's farm (e.g. we held 42 pasture walks on 42 different farms and 30 winter pasture management meetings during 1994). Meetings concern ration balancing, cow body condition scoring and importance, dairy farm enterprise analyses, all aspects of pasture management, grazing behavior, livestock parasite control, seasonal dairying, high through-put milking parlor design, Cabin Fever, and anything else suggested by group members.

In July 1996, farmers participating in the support program formed the Vermont Grass Farmers Association, which now (1997) has 170 members. Its mission is to continue promoting pasture-based livestock production, by assisting individuals and focused discussion groups on grazing issues, and by working with Extension to conduct pasture walks during the grazing season and workshops and conferences during the winter.

By cooperation and sharing experience, each farmer becomes an agent of change through a ripple effect that results around each farm. In this way pasture management information and technology is being transferred quickly and efficiently from farmer to farmer. Discussion group members have progressively come to trust each other, and are willing to share experiences and discuss their problems. Clearly, these have become important support groups for the individual farmers (see economic study in chapter 11).

The Kellogg Foundation evaluation of our grazier support program stated: "Sustainable agriculture is a vital part of the future of rural Vermont. While "sustainable agriculture" has been endorsed by many organizations, including the University of Vermont, there has been little substance behind the endorsement. The grazier support network goes beyond talk to actually implementing an innovative program on real farms run by real farmers. When asked about the most important part of the pasture program, most participating farmers talked about quality-of-life improvements rather than herd health, grain bills, cash flow, or their bottom line. The program serves as a model of grass-roots diffusion of technology rather than the more traditional top-down approach."

CABIN FEVER
In Vermont there is a curious illness among farmers called Cabin Fever. It is joked about, including in books concerning the quaintness of Vermonters that tourists buy, hoping to take something essential of Vermont home with them to tide them over until their next visit. Like many social problems, Cabin Fever is almost never discussed seriously.

Unfortunately, Cabin Fever is no joking matter. It is anxiety and depression, and it affects and endangers many farmers in Vermont and elsewhere.

In the emphasis to achieve maximum production per animal or per acre in fence-row-to-fence-row, get-big-or-get-out, damn-the-cost farming that has been pushed by government, university, and agribusiness, the well being of farmers and farm families has been overlooked. For your sake as a farmer, and our sake as a nation, let's look at it a little bit here.

Farming is one of the best professions that exists. It is absolutely essential to sustain everything else in society. You are your own boss, completely independent, well.... almost. But it's one of the loneliest professions: you work alone. In the past when there was a strong sense of community and neighbors helped and visited each other frequently, working alone was less of a problem, because of the social interaction and support.

Many more people farmed in those days, and farmers were more appreciated. Those were the days before cash flow, debt loads per cow, leveraging the farm, and all the stress and tension that these things have brought with them.

Farming has always been hard, tedious work. It has gone on every day, including weekends and holidays. It has not been a 5-day-a-week, 9-to-5 job. This may surprise the 98 percent of the American population who don't farm, but depend absolutely on the willingness and ability of farmers to continue farming and supporting life in the cities, suburbs, towns, and villages of this country.

"I guess if I felt appreciated by city folks, I could put up with all the problems and the work load," a young farmer told me last summer, going right to the heart of the matter. I think that the greatest problem facing farmers is not herd average, corn yield, or the bottom line; it's the sense of isolation and lack of appreciation that they feel.

For months at a time, the only people that a farmer may talk to from off the farm are the milk truck driver, milk inspector, veterinarian, feed salesman, artificial inseminator, traveling farm supply salesman, and equipment dealer. I wonder how many unnecessary things are bought just so salesmen will return to the farm and provide some company and social interaction for the farmer?

In a world of real values, farmers would be honored by the society they support, at least as much as it honors actors and athletes. Without farmers, the entertainment they provide wouldn't be worth much.

I think that we should create a national holiday to honor farmers: Farmer Appreciation Day. We have days honoring every other group imaginable; why not honor the people who make this country possible? There could be parades and award shows during prime-time TV on Farmer Appreciation Day.

State or federal government could create, train, and subsidize a Farm Work Relief Corps to do farm chores on every farm for a month, so every farm family could have an annual vacation. The Swiss do this with relief milkers, because they value farmers; they know that without farmers Switzerland wouldn't exist. These kinds of activities not only would be extremely helpful to farmers, it would educate city folk about what it takes to provide them with their dependable high-quality food supply.

What To Do About Cabin Fever

A few winters ago we scheduled a discussion of Holistic Management for group meetings during February, right when Cabin Fever is at its peak. At this time, last summer is only a faint memory, Mud Season is fast approaching, and it's just plain damn cold and bleak in Vermont. In our newsletter about the meetings, Lisa McCrory (my associate who wrote the letter) stated that I would concentrate on helping farmers to find the "weak link" in their farms.

I worried and lost sleep about that for 2 months before the meetings, because I knew from my experience of working closely with farmers and listening to what they were saying, what the weak link was. It wasn't milk yield or price, pasture plant varieties, soil fertility, animal nutrition, or any other easy problem to deal with: it was the emotional problems of the farmer him/herself.

During that worrisome time I came across two books called "Feeling Good: The New Mood Therapy" and "The Feeling Good Handbook" by Dr. David Burns (see References). In the books Dr. Burns discusses how to deal with and cure low self-esteem, anger, anxiety, guilt, pessimism, and procrastination -- all characteristics of Cabin Fever depression.

According to Dr. Burns, psychiatrists now think that mainly our thoughts cause and affect our feelings, moods, motivation, and relationships with others. In other words Cabin Fever results from incorrect or twisted ways people think about and interpret things in their situation. Usually this is an habitual way of thinking or behaving that people learn from other people, such as their parents. Since it's due to a *learned* way of thinking, people can break the habit and train themselves to think differently, in a way that doesn't lead to depression. The technique is called cognitive therapy (i.e. thought therapy).

At the risk of turning people off and offending them, I decided to discuss cognitive therapy at the group meetings to determine if depression is the weak link on farms, and to cure Cabin Fever.

To my relief, people attending the meetings said they were glad that depression was recognized as being an important and difficult problem for them. They were relieved that it finally could be discussed and dealt with.

I'm discussing this here, because of that response from those people, and because Dr. Burns's books have greatly helped me. I had Cabin Fever for a long time, but am over it now, thanks to this method and changes in my diet (discussed at end of this chapter). I was amazed and relieved that other people had the same problems as me. Fortunately, I was able to learn how to cure myself.

Some people may wonder what in the world this has to do with managing pasture, but it's clear that we can't do anything well if we're adversely affected by emotional problems. These problems are just as, if not more important as paddock size, pregrazing mass, and recovery period. They need to be resolved right along with the pasture management problems.

So, what follows is information about habitual twisted thinking that can be changed, and two tests from Dr. Burns' books to help you determine if you are affected by anxiety or depression. If you have a problem, depending on how severe it is, you can help resolve it by applying the principles and techniques described in the books.

Forms Of Twisted Thinking That Can Be Changed

1. **All-or-nothing thinking:** You see things in black-or-white categories. If a situation falls short of perfect, you see it as a total failure.

2. **Overgeneralization:** You see a single negative event, such as a cow dying, as a never-ending pattern of defeat and use words such as "always" or "never" when you think or talk about it.

3. **Mental filter:** You pick out a single negative detail and dwell on it exclusively, so that your vision of all reality becomes darkened, like the drop of ink that discolors a beaker of water.

4. **Discounting the positive:** You reject positive experiences by insisting they "don't count." Discounting the positive takes the joy out of life and makes you feel inadequate and unrewarded.

5. **Jumping to conclusions:** You interpret things negatively when there are no facts to support your conclusion. Mind reading: without checking it out, you arbitrarily conclude that someone is reacting negatively to you.

6. **Magnification:** You exaggerate the importance of your

problems and shortcomings, or you minimize the importance of your desirable qualities.

7. **Emotional reasoning**: You assume that your negative emotions necessarily reflect the way things really are.

8. **"Should statements"**: You tell yourself that things *should* be the way you hoped or expected them to be. "Should statements" directed against yourself lead to guilt and frustration. Should statements directed against other people or the world in general lead to anger and frustration. Many people try to motivate themselves with shoulds and shouldn'ts, as if they were delinquents who had to be punished before they could be expected to do anything. This usually doesn't work because all these shoulds and musts make you feel rebellious and you get the urge to do just the opposite.

9. **Labeling**: Labeling is an extreme form of all-or-nothing thinking. Instead of saying "I made a mistake," you attach a negative label to yourself, such as "I'm a loser" or "I'm a failure." Labeling is quite irrational because you are not the same as what you do. Labels are useless abstractions that lead to anger, anxiety, frustration, and low self-esteem. Labeling other people makes you feel hostile and hopeless about improving things and leaves little room for constructive communication.

10. **Personalization and blame**: Personalization occurs when you hold yourself personally responsible for an event that isn't entirely under your control. Some people do the opposite. They blame other people or their circumstances for their problems, and they overlook ways that they might be contributing to the problem.

Tests For Determining Degree Of Cabin Fever
Below are two tests developed by Dr. Burns to help you measure your level of anxiety and depression. Go through

the symptoms and score each one. When you're finished, add up your scores and compare them to the appropriate table. As you work on changing the way you think and feel, you can take these tests every week and compare your scores to monitor your progress to a better quality of life.

Score symptoms according to how much they affect or bother you:

0 = Not at all.
1 = Somewhat.
2 = Moderately.
3 = A lot.

The Burns Anxiety Inventory

Following is a list of symptoms that people sometimes have. Score each one according to how much that symptom or problem has bothered you during the past week.

Category I: Anxious Feelings

1. Anxiety, nervousness, worry, or fear.
2. Feeling that things around you are strange, unreal, or foggy.
3. Feeling detached from all or part of your body.
4. Sudden unexpected panic spells.
5. Apprehension or a sense of impending doom.
6. Feeling tense, stressed, "uptight," or on edge.

Category II: Anxious Thoughts

7. Difficulty concentrating.
8. Racing thoughts or having your mind jump from one thing to the next.
9. Frightening fantasies or daydreams.
10. Feeling that you're on the verge of losing control.
11. Fears of cracking up or going crazy.
12. Fears of fainting or passing out.
13. Fears of physical illnesses or heart attacks or dying.
14. Concerns about looking foolish or inadequate in front of others.

15. Fears of being alone, isolated, or abandoned.
16. Fears of criticism or disapproval.
17. Fears that something terrible is about to happen.

Category III. Physical Symptoms
18. Skipping, racing, or pounding of the heart (sometimes called "palpitations."
19. Pain, pressure, or tightness in the chest.
20. Tingling or numbness in the toes or fingers.
21. Butterflies or discomfort in the stomach.
22. Constipation or diarrhea.
23. Restlessness or jumpiness.
24. Tight, tense muscles.
25. Sweating not brought on by heat.
26. A lump in the throat.
27. Trembling or shaking.
28. Rubbery or "jelly" legs.
29. Feeling dizzy, lightheaded, or off balance.
30. Choking, smothering sensation, or difficulty breathing.
31. Headaches or pains in the neck or back.
32. Hot flashes or cold chills.
33. Feeling tired, weak, or easily exhausted.

Total Score	Degree of Anxiety
0-4	Minimal or no anxiety
5-10	Borderline anxiety
11-20	Mild anxiety
21-30	Moderate anxiety
31-50	Severe anxiety
51-99	Extreme anxiety or panic

The Burns Depression Checklist

Score how much each of these feelings has been bothering you in the past several days. Make sure that you answer all the questions. If you feel unsure about any, go with your best guess.

1. Sadness: Have you been feeling sad or down in the dumps?
2. Discouragement: Does the future look hopeless?
3. Low self-esteem: Do you feel worthless or think of yourself as a failure?
4. Inferiority: Do you feel inadequate or inferior to others?
5. Guilt: Do you get self-critical and blame yourself for everything?
6. Indecisiveness: Do you have trouble making up your mind about things?
7. Irritability and frustration: Have you been feeling resentful and angry a good deal of the time?
8. Loss of interest in life: Have you lost interest in your career, your hobbies, your family, or your friends?
9. Loss of motivation: Do you feel overwhelmed and have to push yourself hard to do things?
10. Poor self-image: Do you think you're looking old or unattractive?
11. Appetite changes: Have you lost your appetite? Or do you overeat or binge compulsively?
12. Sleep changes: Do you suffer from insomnia and find it hard to get a good night's sleep? Or are you excessively tired and sleeping too much?
13. Loss of libido: Have you lost your interest in sex?
14. Hypochondriasis: Do you worry a great deal about your health?
15. Suicidal impulses: Do you have thoughts that life is not worth living or think that you might be better off dead? (Anyone with suicidal urges should seek immediate help with a qualified psychiatrist or psychologist.)

Total Score	Degree of Depression
0-4	Minimal or no depression
5-10	Normal but unhappy
11-20	Borderline-mild depression
21-30	Moderate depression
31-45	Severe depression: seek help with a qualified psychiatrist psychologist.

--∞--

Dr. Burns's books helped me relieve much of the Cabin Fever that I had had all my life. The final cure came by chance when I watched two public television programs called "Spontaneous Healing" and "Eight Weeks to Optimum Health" by Dr. Andrew Weil. What he said made so much sense that I bought and read his books.

From Dr. Weil's books, I realized that besides my habitual twisted thinking, my addiction to caffeine was causing me to feel anxious and depressed, and I needed to get caffeine out of my life. I had been addicted to caffeine for at least 40 years and didn't like the thought of giving up that delicious coffee, tea, and occasional soft drink. I especially dreaded the 3-day withdrawal headache that I would get if I quit drinking them. In the end, it was easy to quit. I switched from coffee to green tea for a month, and then quit drinking the tea: almost no headache!

To my relief, I don't crave caffeinated beverages, and I certainly don't miss the ups and downs (especially the downs) of the usual caffeine addict's day. Cabin Fever disappeared! Getting rid of caffeine also has eliminated the early signs of prostate problems (if you have them, you know what I mean).

Just as in pasture management, where problems are best avoided by good planning and preventive action, I agree with Dr. Weil that it's better to maintain good health by taking preventive measures, rather than trying to cure problems later.

According to Dr. Weil, our bodies are like continuously flowing streams. We can put a certain amount of junk and garbage into them and they will clean themselves, carrying the pollution out with the flow. Overload them with too much pollutants for too long and they become slow and sluggish, depositing things such as fat and cigarette tar in the most unbecoming and dangerous places that can result in severe, complicated health problems. Fortunately, if we stop putting pollutants into our bodies they, like streams, eventually clean and restore themselves to good health.

For example, just as in cows, excess protein must be eliminated from our bodies. This stresses the liver and kidneys, and energy is used to excrete it. In humans it gets more complicated because calcium is thrown out with the excess protein. Many people who suffer from calcium loss think that they need to eat foods richer in calcium, when they may only need to decrease their protein intake.

Following Dr. Weil's advice, I have greatly reduced my intake of fat and protein, and take the vitamins and antioxidants (vitamins B-complex, C, and E, beta carotene, and selenium) and herbal tonics (ginseng and ginkgo biloba) that he recommends. Vitamin B-6 and ginkgo biloba reportedly help reduce feelings of depression. Also, since it is unknown what pesticide residues in food do to our bodies -- it can't be good! -- I eat organically grown food whenever possible. I feel wonderful -- better than ever!

Try these things and see if they help you. I wish you well!

REFERENCES & RECOMMENDED BOOKS

Burns, D.D. 1980. *Feeling Good: The New Mood Therapy.* Avon Books. ISBN 0-380-71803-0. $5.00

Burns, D.D. 1989. *The Feeling Good Handbook.* Avon Books. ISBN 0-452-26174-0. $13.00.

Weil, A. 1995. *Natural Health, Natural Medicine.* Houghton-Mifflin. ISBN 0-395-73099-6. $13.00.

Weil, A. 1995. *Spontaneous Healing.* Fawcett Columbine. ISBN 0-449-91064-4. $12.95.

13
Grazing Disadvantages

"Grazing isn't for everyone."
Strange, I've never heard that kind of apology
Stated about unnecessary year-round
Confinement feeding.

O'Murphy

Some people who either aren't farming, feed in confinement, and/or have some weird vested interest in farmers continuing to feed in confinement have accused me of being one-sided about feeding livestock on well-managed pasture. Whenever they discuss anything about feeding livestock on pasture, they *always* insert the disclaimer or apology that grazing isn't for everyone. For some reason, they haven't felt the same need to apologize for year-round confinement feeding (a little one-sided?).

The people who have accused me of being one-sided say that Voisin management intensive grazing just can't be as simple, profitable, and wonderful an experience as I would have people believe. *That* isn't true; just read the back cover of this book. I tell farmers right out back not to believe me about it! But the critics are right; I have been one-sided in presenting the advantages of pasturing livestock. So this chapter is about some disadvantages.

I tried, but couldn't think of a better description of grazing disadvantages than what Jim Brown of Yuba, Wisconsin wrote to the Editor of *Agri-View* (Jan. 1993):

"I'd like to respond to the person who asked about the negatives, or disadvantages, to grazing. Here's a few I've encountered; these are all things I don't get to do much of anymore.

I don't get to ride a big powerful tractor, and listen to the roar of exhaust. I don't get to see the fuel delivery man much anymore, either. In fact, I only used 1,000 gallons this year for a 60-cow herd, or less than 20 gallons per cow. Before, I was using 50 gallons per cow.

I don't get to visit with the vet much anymore: an $800 (per year) vet bill doesn't bring them out too often.

I don't get to green chop every day like before, or play ring-around the green feed rack, slippin' and sliding in the you-know-what. Speaking of "what," I don't get to haul much during the grazing season; the cows deposit it in the paddocks. That, in turn, makes it so I don't get to buy fertilizer because they poop and pee 50 pounds a day. Two hundred days of grazing makes 10,000 pounds of the stuff. It takes about one acre per cow, so 5 tons ends up being a lot of fertilizer per acre.

It isn't feasible to use a large diesel tractor, so I "have to have" a small gas tractor (the only luxury it's been necessary to buy). A John Deere B doesn't get many "wows" so I have to do without an inflated ego. But I have to admit it's enjoyable to putt around dragging and clipping paddocks on it.

I can only graze the 200 days from April 15 to November 1, so I only get to feed out of storage for 150 days. Also I don't get to formulate fancy rations with a feed consultant (they used to be called salesmen). You see, the only feed the cows receive is about 8 pounds of ground corn a day for the 200 days they're on pasture. And I don't get to do much recreational tillage, as I don't grow very much corn now.

Since I started grazing, I haven't been able to claim a loss on my income tax. I think another reason is, I haven't entered the "bragging contest" of high herd averages. And I don't get to buy any protein supplements anymore.

I don't raise a lot of young stock. The cows seem to last forever, consequently I don't get to break in many new heifers to milking.

Also, I don't get to throw a switch to feed my cows. I have to walk to the pasture and observe, and think, how much to give them today, and tomorrow. And answer 1,000 questions like, "Grandpa what kind of bird is that?" and "Wow, look at this bug!"

Keep those articles on grazing coming. They're appreciated by someone who's grazing, and doing without."

Conclusion

Seek the message,
Not the messenger

Unanimous

The evidence is overwhelming: Voisin management intensive grazing works and it pays! Don't accept pastures for what they are any longer. Think of what they could be if you would manage them with the same amount of attention that you give to your other cropland.

Feeding your livestock on well-managed pasture can give you the time, energy, and money to enjoy life more, while knowing that what you are doing is absolutely essential to society and the Earth. Isn't *that* what farming is all about?

So, do these things, then lay back and enjoy your healthier, happier, more profitable farm, with its greener pastures on your side of the fence!

Appendix

RECOMMENDED READING/VIEWING

Graziers' Monthly Newspaper
The Stockman Grass Farmer, edited by Allan Nation. Subscription: $28/year for US, Canada, & Mexico; $60/year elsewhere; PO Box 9607, Jackson, MS 39286-9607. Phone: 800-748-9808; Fax: 601-981-8558.

Books

Available from The Stockman Grass Farmer and other book sellers
Dollars and Sense: A Handbook for Seasonal Grass Dairying by Larry Tranel. $17.95.

Grass Farmers by Allan Nation. $23.50.

Grass Productivity by Andre Voisin. Island Press 1988 reprint. $22.

Grazier's Resource Guide. Free.

Intensive Grazing Management by Burt Smith. $29.95.

Pastured Poultry Profits by Joel Salatin. $30.

Pasture Profits with Stocker Cattle by Allan Nation. $24.95.

Quality Pasture by Allan Nation. $32.50

Salad Bar Beef by Joel Salatin. $30.

Shelter and Shade by John & Bunny Mortimer. $20.

Available from publisher and other book sellers

Grasses by Lauren Brown. $10.95. Houghton Mifflin Co.

Holistic Resource Management by Allan Savory. $25. Center for Holistic Management, 1010 Tijeras NW, Albuquerque, NM 87102. Phone: 505-842-5252; Fax: 505-843-7900.

Weeds by Walter Muenscher. $22.50. 1987 reprint. Cornell University Press. First published in 1960 before general use of herbicides, it describes cultural methods of controlling weeds.

Weeds of the Northeast by Richard Uva, Joseph Neal, & Joseph Diomaso. $25. Cornell University Press. Provides color photographs that make identification easier.

Videos

Available from The Stockman Grass Farmer
Grass Dairying by John Cockrell. 30 min. $24.95.

Milking Systems -- New Zealand Style by Vaughn Jones. 17 min. $19.95.

Pastured Poultry Profits by Joel Salatin. 45 min. $50.

The Making & Feeding of Silage by Vaughn Jones. 45 min. $34.50.

NOXIOUS WEEDS

Name, life cycle, reproduction, cultural controls	Harmful part

Mechanical or chemical injury

Bedstraw (*Galium* species)   Tops
Perennial; seeds & rootstocks; mow early before seeds are produced; close grazing in early spring prevents it from spreading.

Bull thistle (*Cirsium vulgare*)   Spines
Bienniel; seeds; cut the rosettes below the crown early in spring; cut repeatedly to prevent flowering and seed set; allow no thistles to produce seed.

Canada thistle (*Cirsium arvense*)   Spines
Perennial; seeds & creeping roots; small patches can be destroyed by digging out the roots; mow repeatedly before plants blossom; mow again if new growth appears.

Downy bromegrass (*Bromus tectorum*)   Awns
Annual; seeds; Mow before seeds form.

Poison ivy, sumac (*Rhus* species)   Leaves, bark: touch
Perennial; seeds & creeping rootstocks; small areas on stony soil can be controlled by close grazing with sheep or goats; grub out rootstocks of scattered plants; mow large areas closely in midsummer, followed by plowing & harrowing, then in spring seed a cultivated crop. Don't touch any part of the plants with skin or clothing and don't breath smoke if you burn them.

Puncture vine (*Tribulus terrestris*)   Floral carpels
Annual; seeds; hoe or cut off below the crown, pile and burn plants before seeds mature.

Russian thistle (*Salsola kali*)   Spine-like leaves
Annual; seeds; mow before seed set; clean cultivation followed by hand
hoeing to kill stray plants; pile and burn mature plants to prevent seeds
from scattering.

Sandbur (*Cenchrus longispinus*)   Spiny burs
Annual; seeds; hoe or pull plants before seeds mature; burn plants with
mature burs.

Squirrel tail grass (*Hordeum jubatum*)   Awns
Biennial or perennial; seeds; graze closely in early spring; plow deeply
and follow with an annual smother crop (e.g. small grain).

Stinging nettle (*Urtica dioica & gracilis*)   Hairs
Perennial; seeds & creeping rootstocks; mow closely to prevent seed
formation; grub rootstocks out & kill by drying.

Wild oats (*Avena fatua*)   Awns
Annual; seeds; Mow in dough stage to prevent reseeding.

Wild parsnip (*Pastinaca sativa*)   Leaves
Biennial; seeds; cut off plant rosettes below soil surface.

Poisonous

American, false hellebore (*Veratum viride*)   Tops & roots
Perennial; seeds & short rootstocks; improve drainage, followed by
cultivation; dig out scattered plants or patches of plants.

Black & white swallow wort (*Cynanchum* spp) Tops
Perennial; seeds; mow closely several times during summer.

Bouncing Bet (*Saponaria officinalis*)   Entire plant
Perennial; seeds & rootstocks; dig out scattered plants; mow large areas
closely every time the first flowers open.

Brake fern (*Pteridium aquilinum*)   Leaves, stems
Perennial; spores & creeping rootstocks; clean cultivation destroys
rootstocks; mow or pull fronds twice a year before spores mature (July-
August); fertilize infested areas and graze closely with sheep.

Buttercup, cursed crowfoot (*Ranunculus* species) Fresh tops
Perennial; seeds; improve drainage; badly infested areas should be
plowed & followed by clean cultivation for a year; fertilize & harrow
pasture in spring; mow buttercup clumps when first flowers bloom &
again when second flowers bloom.

Celandine (*Chelidonium majus*)   Tops
Biennial; seeds; hand pull or hoe; clean cultivate.

Choke cherry (*Prunus virginiana*)   Wilted leaves
Perennial; seeds & new shoots from roots; dig out bushes; cut in midsummer & burn.

Cocklebur (*Xanthium orientale*)   Seeds, seedlings
Annual; seeds; mow before seeds form; burn mature plants.

Cow cockle (*Saponaria vaccaria*)   Entire plant
Annual; seeds; pull out plants.

Darnel (*Lolium temulentum*)   Seeds with fungus
Annual; seeds; include a cultivated crop in a short rotation.

Foxglove (*Digitalis purpurea*)   Leaves & stem
Biennial; seeds; cut off plant rosettes in early spring; mow closely when flower stalks appear.

Horse nettle, European bittersweet, Black
nightshade, Buffalo bur (*Solanum* species)   Tops & berries
Perennial (black nightshade is annual); seeds & creeping rootstocks; mow scattered patches frequently; clean cultivation of large areas.

Horsetail, scouring rush (*Equisetum* species)   Tops
Perennial; spores & rootstocks; improve drainage & soil fertility to favor pasture plants; frequent close mowing; cleanly cultivate.

Indian tobacco (*Lobelia inflata*)   Tops
Annual; seeds; cut or pull out plants; mow closely in spring and again before seeds mature.

Jimsonweed (*Datura stramonium*)   Leaves, fruit, seeds
Annual; seeds; mow before seeds form; burn plants with mature seedpods. Don't touch plants with bare skin and don't breath smoke.

Larkspur (*Delphinium menziesii & stritum*)   Tops
Perennial; seeds & roots; dig out plants & roots when they begin to blossom; graze with sheep.

Laurel (*Kalmia* species)   Leaves
Perennial; seeds & new shoots from roots; dig out roots; mow or cut tops & burn.

Lupines (*Lupinus* species)   Tops & seed pods
Perennial; seeds & rootstocks; pull or hoe scattered plants.

Meadow death camas (*Zigadenus venenosus*)   Tops & bulbs
Perennial; seeds & bulbs; dig bulbs when soil is wet, & destroy them.

Oak (*Quercus* species)   All edible parts
Perennial; acorns; fence livestock away from trees.

Poison hemlock (*Conium maculatum*)   Tops & roots
Bienniel; seeds; cut off plant rosettes below soil surface.

Pokeweed (*Phytolacca americana*)   Entire plant, esp. roots
Perennial; seeds; plow badly infested grasslands & seed to cultivated
crop for 1-2 years; cut scatteredplants below the crown.

Rattlebox (*Crotolaria sagittalis*)   Tops
Annual; seeds; clean cultivation.

Scarlet pimpernel (*Anagallis arvensis*)   Tops
Annual; seeds; pull or hoe scattered plants before first blossoms appear.

Sneezeweed, bitterweed (*Helenium* species)   Leaves & stem
Perennial; seeds; improve drainage; pull or mow plants as soon as
flowers appear.

Spurges (*Euphorbia* species)   Tops
Perennial; seeds & rootstocks; dig scattered plants; mow closely before
blossoms appear.

St. Johnswort (*Hypericum perforatum*)   Tops
Perennial; seeds & short runners; mow several times to prevent seed
formation.

Star of Bethlehem (*Ornithogalum umbellatum*) Entire plant
Perennial; bulbs; dig out plants & dry or burn bulbs.

Velvet grass (*Holcus lanatus*)   Leaves, stems
Perennial; seeds; mow closely early before seeds are produced; plow and
follow with a smother crop (e.g. small grain) or a clean-cultivated crop
for 1 or 2 years before reseeding.

Water hemlock (*Cicuta maculata*)   Tops & roots
Perennial; seeds & roots; pull or dig plants early in spring while soil is
soft; mow before seeds form.

White snakeroot (*Eupatorium rugosum*)   Leaves
Perennial; seeds; mow closely several times before seeds form; improve
drainage.

Wild mustard (*Brassica kaber*)   Seeds
Annual; seeds; hand pull when plants begin to blossom.

Produce undesirable flavors in dairy products

Fanweed, pennycress (*Thlaspi arvense*)   Leaves, stems
Annual; seeds; mow when blossoms start to form.

Marsh elder, false ragweed (*Iva axillaris*)   Leaves, stems
Annual; seeds; mow before flowers open.

Mustard (*Brassica* species)   Leaves, stems
Annual; seeds; pull scattered plants as soon as they begin to blossom; mow large areas when blossoms start to form.

Ragweed (*Ambrosia* species)   Leaves, stems
Annual; seeds; mow before flowers open.

Wild carrot, Queen Anne's lace (*Daucus carota*) Tops
Biennial; seeds; mow as soon as plants begin to blossom.

Wild garlic (*Allium vineale*)   Tops
Perennial; bulbs & bulblets; dig scattered clumps or plants; for large areas, disk to destroy tops, then plow in late autumn, follow by early spring plowing and seed a clean cultivated crop.

Wild leek (*Allium tricoccum*)   Tops
Perennial; seeds & bulbs; dig out bulbs as soon as leaves appear.

Wild onion (*Allium canadense*)   Tops
Perennial; bulbs, bulblets, & seeds; dig out bulbs before bulblets form; for large areas plow in autumn, follow with a clean cultivated crop for 1-2 years before reseeding to pasture.

Adapted from Muenscher, W.C. *Weeds*. 1987 reprint, Cornell University Press.

Index

ORDER FORM

Arriba Publishing
2238 Middle Road
Colchester, VT 05446
Telephone 800/639-4178

Please send me Greener Pastures On Your Side Of The Fence: Better Farming With Voisin Management Intensive Grazing (4th Edition) by Bill Murphy @ $30.00.

I understand that I may return the book for a full refund if I'm not satisfied.

Enclosed for   book(s) is   _____

(Vermonters please add 5% sales tax: $1.50/book) _____

Packing and shipping:

$4.50 per book   _____

(Multiple orders cost less per book
to ship and will be billed accordingly.)

Total enclosed   _____

Name: _____

Address: _____

_____

ORDER FORM

Arriba Publishing
2238 Middle Road
Colchester, VT 05446
Telephone 800/639-4178

Please send me Greener Pastures On Your Side Of The Fence: Better Farming With Voisin Management Intensive Grazing (4th Edition) by Bill Murphy @ $30.00.

I understand that I may return the book for a full refund if I'm not satisfied.

Enclosed for   book(s) is   _____

(Vermonters please add 5% sales tax: $1.50/book) _____

Packing and shipping:

     $4.50 per book   _____

     (Multiple orders cost less per book to ship and will be billed accordingly.)

            Total enclosed   _____

Name: _____

Address: _____

_____